T0336398

Hadronic Jets (Second Edition)

An introduction

Online at: https://doi.org/10.1088/978-0-7503-4737-2

Hadronic Jets (Second Edition)

An introduction

Andrea Banfi

School of Mathematical and Physical Sciences, University of Sussex, Brighton, UK

IOP Publishing, Bristol, UK

ISBN 978-0-7503-4737-2 (ebook)
ISBN 978-0-7503-4735-8 (print)
ISBN 978-0-7503-4738-9 (myPrint)
ISBN 978-0-7503-4736-5 (mobi)

DOI 10.1088/978-0-7503-4737-2

Version: 20221201

IOP ebooks

British Library Cataloguing-in-Publication Data: A catalogue record for this book is available from the British Library.

Published by IOP Publishing, wholly owned by The Institute of Physics, London

IOP Publishing, No.2 The Distillery, Glassfields, Avon Street, Bristol, BS2 0GR, UK

US Office: IOP Publishing, Inc., 190 North Independence Mall West, Suite 601, Philadelphia, PA 19106, USA

To my PhD students.

Contents

Preface

The physics of hadronic jets is an incredibly rich subject, encompassing not only theoretical and experimental physics, but advanced mathematics and computer science as well. My daily research activity considers jets from a specific angle, that of precision calculations, and has the ambition of understanding complicated observables involving jets by means of analytic calculations. However, even such a focus requires to have a broad overview of the topic, both from a theoretical and experimental point of view. The subject is also constantly evolving, not only due to new ideas, but also as an important development in computing techniques, in particular artificial intelligence. This is the knowledge I have tried to share with the present book, which can be thought of as a gentle introduction to jet physics.

This book aims at presenting the main theoretical and experimental ideas that had made it possible for jet physics to bloom, and to become one of the hot topics in particle physics. It can be then useful to physicists not directly involved in the field, in the hope that, seeing how relevant problems in jet physics have been solved, they could find valuable inspiration for their own research. Also, many times during my career I have met students, both graduate and undergraduate, willing to start a project on jet physics, but without an extensive background in quantum field theory. Such students are not ready to understand a topical review on the subject, or a book on collider physics. These are the second kind of readers I had in mind. This is why, instead of starting from first principles, I have decided to privilege basic experimental facts, and work from there to present their theoretical interpretation. This of course requires performing some calculations. In presenting them, in this second edition I focused even more on introducing simple fully worked-out examples, that nevertheless encompass the main theoretical ideas underlying the state of the art. Last but not least, I thought about the book as a pleasant read, where details could eventually be skipped, and in which each chapter could be read independently from the others. The basic organisation of the book has not changed from the previous edition, although some of the content has been updated. The introduction aims at explaining what hadronic jets are, and why they are so important for our current understanding of the physics of elementary particles. This chapter also contains an express review of particle physics, so that the reader gets used to the language employed in the rest of the book. This is followed by a chapter on jet algorithms, that describes the procedures that are currently adopted to rigorously define jets. With respect to the first edition, this chapter contains an update on the procedures used to mitigate the effect of pile-up and underlying event, which nowadays makes massive use of machine learning techniques. Also, the final section on novel jet algorithms has been updated, both in shortening the discussion on promising procedures that, however, did not have the expected impact, and in discussing in more depth a suggestive geometrical interpretation of jet algorithms. The first two chapters can be understood by a reader with a solid background in fundamental physics, with no detailed knowledge of particle physics required. Chapter 4 is devoted to one important topic in high-energy physics, the search for new particles that decay into jets. With respect to the first edition, I have given some space to the onset of artificial intelligence, which is now a widely used tool for such studies. Chapter 4 could, in principle, be understood with the material contained in chapter 2, should the reader skip some basic calculations. However, the main ideas presented in chapter 4 can be better appreciated by a reader familiar with quantum chromodynamics (QCD), the quantum field theory that provides the theoretical foundations of jet physics. Therefore, I have decided to devote chapter 3 to presenting QCD as the origin of the main theoretical tools that are nowadays used to describe jets. Chapter 3 is not an extensive presentation of QCD, but rather an illustration of how QCD is used in jet physics. There, every theoretical idea is presented through an example, followed by a review of how the same idea is actually implemented in current theoretical tools. It is my hope that the reader might understand the tasks that each tool actually performs, and to which physical situations it can be reliably applied.

Each chapter contains its own list of references, by no means complete. These are the ones I would suggest an interested reader to go through, so as to have a deeper understanding of the covered topics. Also, the book expresses a personal view on the subject, so I felt free to select which results to present. In fact, in order to help the book flow, I had to sacrifice an important topic like the production of jets with wide angular gaps between them, and relevant concepts like soft-gluon interference, renormalisation and parton density functions are only quickly mentioned.

Lastly, the book has been developed during the runs of the Large Hadron Collider at CERN. This is the place where many of the methods presented here reveal their full potential. I very much hope that groundbreaking discoveries may occur thanks to them.

Acknowledgements

This book elaborates on various lectures for PhD students I delivered in Parma, Freiburg, Cargese, Hamburg and Amsterdam. I am therefore very thankful to the organising committees of those schools, in particular, to Enrico Onofri, Karl Jacobs, Christophe Grojean, Hannes Jung and Eric Laenen for the invitations and for having given me the possibility to share the physics I love most. I am also grateful to all the students I met there for their questions, comments and valuable discussions and suggestions. I am deeply in debt with Pino Marchesini (who sadly passed away in 2016) and Yuri Dokshitzer, two extraordinary physicists who taught me QCD, and supported me throughout my career. I have also benefited much from my longstanding collaboration with Gavin Salam, Giulia Zanderighi and Mrinal Dasgupta, as well as my most recent collaboration with Pier Francesco Monni. I also acknowledge the benefits of a healthy competition on QCD resummations with Thomas Becher.

Much of the material on jet algorithms and jet substructure is the result of many illuminating discussions with Gavin Salam, Matteo Cacciari, Gregory Soyez, and Mrinal Dasgupta. I acknowledge also many fruitful exchanges on the topic with Andrew Larkoski, Simone Marzani, Matthew Schwartz and Jesse Thaler.

The part on fixed-order calculations reflects what I have learnt from experts in the field I met during my postdoc in Zurich. In particular, I acknowledge many discussions with Babis Anastasiou, Achilleas Lazopoulos, Thomas Gehrmann and Massimiliano Grazzini.

I also wish to thank all my colleagues at Sussex for many physics discussions, in particular, Jonas Lindert, Fabrizio Salvatore, Iacopo Vivarelli, who gave me valuable insight on many theoretical and experimental aspects of jet physics at the Large Hadron Collider.

Author biography

Andrea Banfi

Andrea Banfi is a theoretical particle physicist, expert in the development of theoretical frameworks for the calculation of observables involving hadronic jets at high-energy colliders. In particular, he is active in the calculation of Higgs production rates in the presence of hadronic jets at the Large Hadron Collider. He obtained his PhD at the University of Milano in 2002, and then worked as a Research Associate for various institutions in Europe, namely NIKHEF (Amsterdam), the Cavendish Laboratory (Cambridge), the University of Milano-Bicocca, ETH Zurich, and as Assistant Professor by the University of Freiburg. In 2013 he joined the Theoretical Particle Physics group of the University of Sussex, where he is now Professor of Theoretical Physics.

IOP Publishing

Hadronic Jets (Second Edition)

An introduction

Andrea Banfi

Chapter 1

Introduction

One of the most striking phenomena that can be observed in high-energy collisions of elementary particles is the production of highly collimated bunches of particles (see figure 1.1). These objects are known as hadronic jets. The word 'hadronic' refers to the fact that jets are made up of hadrons, particles which can interact through the strong force, the one that keeps atomic nuclei bound together. If we look inside a jet we might find protons and neutrons, the constituents of nuclei, and other less known hadrons such as pions, which are commonly observed as cosmic rays, as well as kaons, rho mesons, etc. Looking at summary tables of hadrons in the '*Review of Particle Physics*' by the Particle Data Group [2], one finds around 50 pages devoted to mesons, hadrons of integer spin (bosons), such as the pions, and another 20 pages devoted to baryons, hadrons of half-integer spin (fermions), such as protons and neutrons. Given such a proliferation of particles, it seems almost a dream to be able to understand anything about jets of hadrons, and even more inconceivable to write a book about them. Surprisingly enough, the main features of hadronic jets, such as the distributions in their energies and angles, have little to do with their constituent hadrons, but rather with the constituents of the hadrons themselves. It is firmly established that hadrons are not elementary particles, but are bound states of point-like particles, the quarks. These are spin-1/2 particles interacting via the strong force, which is mediated by spin-1 gauge bosons, the gluons. Quarks and gluons are commonly referred to as 'partons'. This is the name that was given to quarks the first time they were probed in inelastic electron–proton collisions at SLAC, in view of the fact that they appeared as parts of the proton [3, 4]. In fact, in such collisions, the scaling properties of the angular distribution of scattered electrons were consistent with the fact that these were deflected by point-like particles of spin-1/2, carrying a fraction of the proton energy and charge (hence the name *partons*). Such a behaviour was more pronounced the higher the momentum transferred by the electron in the collision. The interpretation of this experimental result is that the hit partons, when probed at high energy, are not tightly bound inside the proton, otherwise the latter

Figure 1.1. A spectacular event with many jets observed by the ATLAS detector at CERN. The picture is taken from the ATLAS public event display repository [1]. ATLAS Experiment, Copyright 2014 CERN.

would recoil against the electron as a whole. After some time, consensus built up in identifying the partons with the quarks, elementary particles whose existence was hypothesised some years before by Gell-Mann to explain the properties of hadrons, in particular their electric charges and spin [5].

The crucial breakthrough that gave theoretical soundness to the identification of the partons observed at SLAC with the quarks introduces by Gell-Mann was the discovery that, in some quantum field theories, the effective interaction strength between elementary particles decreases with increasing energy of the particles involved [6, 7]. This property, known as 'asymptotic freedom', was assumed to hold for the theory governing the interactions between quarks. Within this framework it is possible to explain inelastic electron–proton collisions. In fact, when the momentum transferred by the electron to the target is small, the quarks interact very strongly and are confined within the proton, which hence interacts with the electron as a whole. At higher momentum transfer, the quarks inside the proton are probed at high energies and they essentially behave as free particles. This picture was consistent with the behaviour of the electron–proton inelastic collisions observed at SLAC.

More specifically, the theory underlying quark interactions is called quantum chromodynamics (QCD), in that the quarks are supposed to carry a new type of charge, called colour ('chroma' in ancient Greek). Quarks interact between each other by exchanging spin-1 particles, called gluons. The latter obtain their name from the fact that they provide the 'glue' that binds quarks together inside hadrons. QCD is quite similar to electromagnetism, with gluons playing the role of photons. The main difference is that gluons carry colour themselves, and therefore can interact directly with other gluons[1].

QCD gives a natural explanation for the occurrence of hadronic jets. In fact, in a high-energy collision, quarks and gluons are abruptly produced, and ripped apart. Each parton (quark or gluon) radiates gluons, very much like an electron smashing on a target radiates x-ray photons. At high energies, the radiated particles are highly collimated in the directions of the original quarks and gluons produced in the primary collision. Through radiation, quarks and gluons degrade their energies, and their interactions become stronger and stronger, until they cluster together to form hadrons, which are the actual 'final-state' particles, i.e. the ones we observe in our detectors. The transformation of partons into hadrons, commonly referred to as the 'hadronisation' process, does not significantly alter the energy–momentum flow of the original quarks and gluons. Therefore, the jettiness of high-energy hadronic events is to be mainly attributed to the properties of radiation of gluons from coloured particles. Jets are thus the footprints of unobservable quarks and gluons in our detectors. Most of their properties can be understood using the language of quarks and gluons, without having to bother with the properties of the individual hadrons that constitute the jets.

These ideas were further confirmed by experiments involving electron–positron collisions. First of all, events with two jets were observed by the SPEAR collaboration at SLAC [8]. The angular distribution of the jets was compatible with the production of a quark–antiquark pair fragmenting in two bunches of collimated hadrons.

The existence of gluons, whose radiation is ultimately responsible for the formation of jets, was confirmed through the observation of three-jet events at the Positron–Electron Tandem Ring Facility (PETRA; Positron–Elektron Tandem Ring Anlage) collider at DESY [9–12].

Such events, one of which is displayed in figure 1.2, could be explained as resulting from radiation of an energetic gluon off a quark–antiquark pair. Subsequent studies of jet distributions in three-jet events were able to assess that the gluon had spin-1, and measurements of angular correlations between jets in four-jet events confirmed the existence of gluon self-interactions, as predicted by QCD (see [13] for a historical overview on the discovery of the gluon and its properties). Jets were also observed in hadronic collisions at the Intersecting Storage Rings (ISR) [14] and at the Super Proton Synchrotron (SPS) at CERN [15, 16], thus providing evidence of scattering between quarks and gluons. These and other results firmly established that QCD is able to successfully account for quark–gluon interactions at high energies, and that jets are the experimental signature that high-energy QCD is at work.

Jet physics is an incredibly rich subject, whose many intriguing aspects are thoroughly discussed in specialised reviews (see e.g. [17]). This book aims at giving a general overview of this topic to non-experts in particle physics, in particular

[1] Photons can interact among themselves as well, but their interaction is always mediated by electrically charged particles.

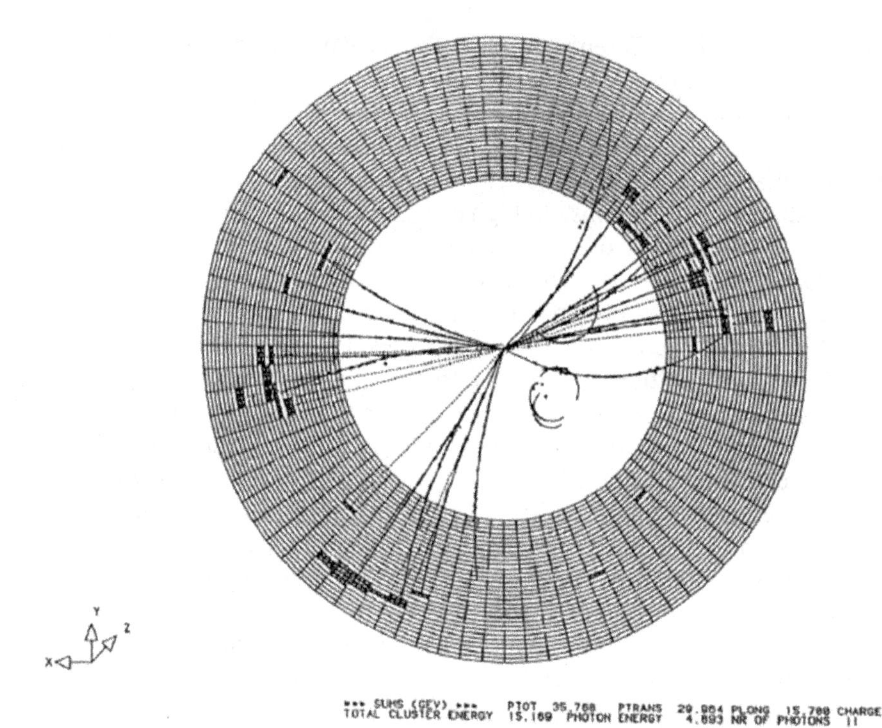

Figure 1.2. A three-jet event as seen by the JADE detector at PETRA. Reproduced from [13] Copyright © 2010, EDP Sciences and Springer, with permission from Springer.

scientists not directly involved in the field, as well as students, both experimentalists and theorists, who intend to start a research project in high-energy physics.

The book is organised as follows. In chapter 2 we discuss jet algorithms, the procedures that are used to rigorously define jets and to extract them from the multitude of hadrons present in a typical final state at high-energy colliders. Chapter 3 will be devoted to QCD, the theory of strong interactions, governing the dynamics of quarks and gluons. In particular, we will describe the theoretical tools within QCD that can be used to describe the properties of jets. Finally, in chapter 4 we will discuss how, from a set of observed jets, it is possible to extract information on the elementary event that has produced them. Such techniques are extremely important for search for new particles, especially when they are expected to decay into quarks and gluons, giving rise to jets in the final state.

The reader who is familiar with elementary particle physics is ready to start with chapter 2. In the following, basic facts on elementary particles and high-energy colliders are presented. These can be considered as the minimal background required to understand the rest of the book.

1.1 Basics of elementary particle physics

The known elementary particles, and their interactions, are organised in the so-called Standard Model of elementary particles, summarised in table 1.1.

Table 1.1. The standard model of elementary particles.

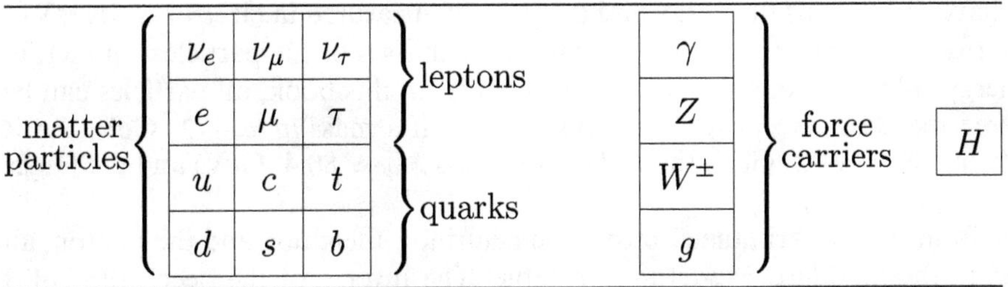

matter particles	ν_e	ν_μ	ν_τ	leptons	γ	force carriers	H
	e	μ	τ		Z		
	u	c	t	quarks	$W^\pm$		
	d	s	b		g		

This represents in a sense the actual table of elements. In the language of relativistic quantum mechanics, each particle is associated with a corresponding quantum field. In fact, the energy of each free propagating field is not a continuous quantity, but is quantised, i.e. made up of elementary excitations. These excitations carry both energy and momentum, and can be interpreted as particles.

The first building block of the Standard Model is matter particles, carrying spin-1/2. These are further divided in three families of leptons and three families of quarks. Each family of leptons is organised into doublets. We have then the electron e with the electron neutrino ν_e, the muon μ and the muon neutrino ν_μ, and the tau τ with the tau neutrino ν_τ. All neutrinos are electrically neutral, whereas the electron, muon and tau have charge $-e$, with $e \simeq 1.6 \times 10^{-19}$ C the magnitude of the charge of the electron (alternatively the charge of the proton). Quark families are also organised into doublets, the first containing the up (u) and down (d) quarks, the second the charm (c) and strange (s) quarks, and the last the top (t) and bottom (b) quarks. All quarks in the top row (up, charm and top) have electric charge 2/3 e, whereas the ones in the bottom row (down, strange and bottom) have charge $-1/3$ e. The fundamental difference between leptons and quarks is that the former are not subject to strong interactions. An important piece of information that is not reported in the table is that each matter particle has a corresponding anti-particle, having the same mass but opposite charge. For instance, the anti-particle of the electron e^- is the anti-electron or 'positron' e^+. Similarly, the anti-particle of a quark q is an antiquark, denoted by $\bar{q}$.

The second part of the table contains force mediators. These are spin-1 particles, whose fields are responsible for the interactions among leptons and quarks. The photon mediates electromagnetic interactions, the W and Z bosons mediate weak interactions, and the gluon mediates strong interactions. All gauge bosons are electrically neutral, except $W^\pm$, whose electric charge is $\pm e$.

The last ingredient of the Standard Model is the Higgs particle, whose field is responsible for giving mass to all particles: the larger the interaction with the Higgs field, the larger the mass of a particle. The Higgs boson is electrically neutral.

Throughout this book, we will use the system of natural units, in which the Planck constant $\hbar$ and the speed of light c are conventionally set to one. In this system of units, the only quantities with dimensions are length, which has the same dimensions as time, and energy, whose dimensions is the inverse of a length. In natural units, all masses have dimensions of energy and are usefully measured in electron-volts (eV).[2] Table 1.1 does not show any information on the masses of the particles. In fact, in high-energy collisions, the regime we will consider in this book, all particles can be considered massless, except for the top quark, with a mass $m_t \simeq 173$ GeV, the Z boson (mass $M_Z \simeq 91.2$ GeV) the W bosons (mass $M_W \simeq 80.4$ GeV) and the Higgs (mass $M_{\mathrm{H}} \simeq 125$ GeV).

Aside from the u quark, the electron, the neutrinos, the gluon and the photon, all particles in the Standard Model are unstable. The inverse of the decay time of a particle in the particle's rest frame is called 'width' and is indicated by Γ. Of course, the larger the width of a particle, the smaller its decay time. In natural units, the width of a particle is measured in electron-volts.

The elementary particles we observe in high-energy experiments have speeds that are very close to the speed of light, so that Lorentz transformations have to be applied to relate quantities measured in one reference frame to another. To simplify equations among the relevant physical quantities appearing in high-energy experiments, it is useful to construct quantities that are invariant with respect to Lorentz transformations. For instance, the energy E and the three-momentum (also known as impulse) $\vec{p}$ of a particle in a given reference frame can be organised in a four-vector $p \equiv (E, \vec{p})$, having well-defined properties under Lorentz transformations. The quantity

$$p^2 \equiv E^2 - \vec{p}^{\,2}, \quad \vec{p}^{\,2} = \vec{p} \cdot \vec{p} = |\vec{p}\,|^2, \tag{1.1}$$

is invariant under Lorentz transformations. Its square root is equal by definition to the mass of the particle. Similarly, if we consider two four-vectors $a = (a^0, \vec{a})$ and $b = (b^0, \vec{b})$, the 'dot' product

$$a \cdot b \equiv (ab) \equiv a^0 b^0 - \vec{a} \cdot \vec{b} \tag{1.2}$$

is also invariant under Lorentz transformations. In order to avoid the complications of performing Lorentz transformations at every corner, it is customary to express all relations between four-momenta (from now on simply 'momenta') in terms of relativistically invariant 'dot' products. For instance, if p_a and p_b are the momenta of the decay products of a particle, one can measure the so-called 'invariant mass' of the decay products, defined as the square root of the invariant

$$(p_a + p_b)^2 = (p_a + p_b) \cdot (p_a + p_b). \tag{1.3}$$

It is possible to show using relativistic quantum mechanics that the distribution in the invariant mass of the decay products of an unstable particle has a peak in correspondence of the actual mass of the particle. The width of the peak, which has in fact the dimensions of an energy, is proportional to the width of the

[2] We recall that one electron-volt (eV) is the work done by the electric force to move an electron between two points whose electric potential difference is one volt.

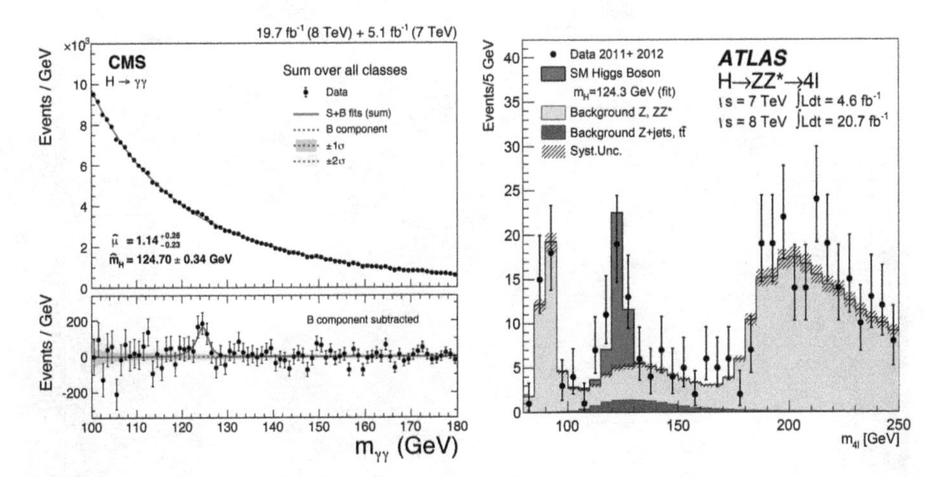

Figure 1.3. Distribution in the invariant mass of two photons ($m_{\gamma\gamma}$) measured by CMS (reproduced from [18] (2014). With permission of Springer) (left), and of two Z bosons (m_{4l}) measured by ATLAS (Reprinted from [19], Copyright (2014), with permission from Elsevier) (right), where a peak corresponding to the production and subsequent decay of a particle with a mass around 125 GeV can clearly be seen.

unstable particle. Looking for peaks in invariant mass distributions is the standard procedure to search for new particles. For instance, the recently discovered Higgs boson appeared first as a peak in the invariant mass of two photons (figure 1.3, left-hand panel), as well as in that of two Z bosons (figure 1.3, right-hand panel).

In this book we will be interested in the properties of elementary particles that can be probed through high-energy collisions. It is therefore useful to spend some time describing how a particle collider works. The first stage is the production and storage of beam particles. These undergo subsequent accelerations until they reach the desired energy. Two beams are then properly focused and made to collide at selected collision points, where suitable detectors have been placed. From the analysis of the signals in the detectors, experimentalists are able to obtain information on the particles produced in each collision. The main quantities of interest at colliders are cross sections, physical observables that are related to the likelihood that events occur, and are independent of the details of the experimental apparatus. More specifically, for a process $ab \to X$, where a and b are the colliding particles, and X a selected final state (for instance, a Higgs boson, plus anything else), the number of observed events per unit time dN_X/dt is related to the cross section $\sigma_{ab\to X}$ via the relation:

$$\frac{dN_X}{dt} = \mathcal{L} \times \sigma_{ab\to X}. \tag{1.4}$$

The quantity $\mathcal{L}$ is called luminosity, and encodes the information on the intensity of the beams, i.e. the rate of incoming particles per unit area. Cross sections have dimensions of an area, and are usually measured in barn (b), with 1 b = 10^{-28} m^2.

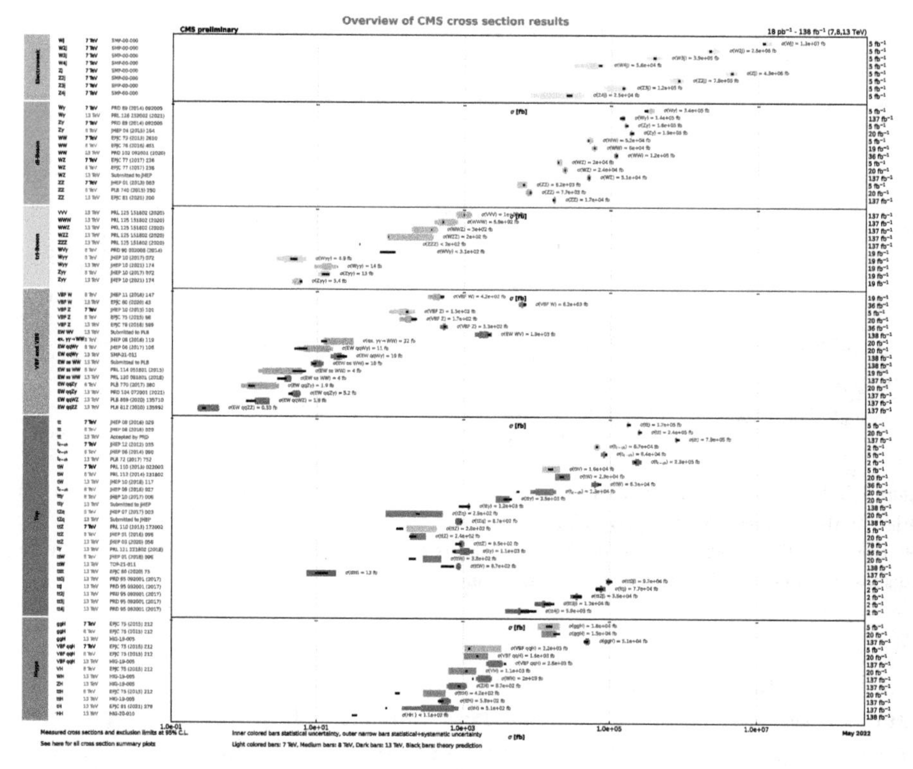

Figure 1.4. Cross sections for selected processes, as measured by the CMS detector during the first runs of the LHC [20]. Copyright &© 2008-2022 by the contributing authors.

One considers also the luminosity accumulated over a certain period of time, the so-called 'integrated luminosity', which is usually measured in b^{-1}. Of course, the higher the integrated luminosity, the better the chances will be of observing rare phenomena. This is illustrated in figure 1.4, where one can see the cross sections for selected processes at the LHC. The main feature to note here is that that with increasing integrated luminosity more processes become observable.

In the following we will describe the different experimental set-ups that can lead to measurements such as the one in figure 1.4. In particular, the two kinds of machines we will consider in this book are electron–positron and hadron–hadron (or simply 'hadron') colliders. Furthermore, we will only deal with experimental set-ups that are relevant for jet physics. In doing this we will not be able to discuss extremely important low-energy machines used to observe particular processes, such as rare hadron decays, or to measure selected physical quantities, such as K^0–$\overline{K^0}$ mixing parameters, with very high precision.

In electron–positron (e^+e^-) colliders, electron and positron beams are accelerated with various techniques and made to collide. Interesting events occur when an electron and a positron from each beam annihilate, and the energy available in the collision gives rise to new particles. The typical configuration of high-energy e^+e^- colliders is that in which the two beams have the same energy E_{beam} and opposite velocities. In this way no energy is wasted in the motion of the centre-of-mass of the system, and a total energy $2E_{\text{beam}}$ is available in the annihilation process. Given the momentum of an electron k_1 and that of a positron k_2, one introduces the relativistically invariant quantity

$$s = (k_1 + k_2)^2, \tag{1.5}$$

so that, in case of two opposite beams with energy E_{beam} each, $\sqrt{s} = 2E_{\text{beam}}$ represents the total centre-of-mass energy available for a collision. Examples of high-energy electron–positron colliders are the PETRA accelerator at DESY, whose outputs were analysed by the JADE [21], MARK-J, PLUTO, TASSO and CELLO [22] experimental collaborations, and the Large Electron–Positron (LEP) collider at CERN, with its four experiments ALEPH [23], DELPHI [24], L3 [25] and OPAL [26]. One advantage of electron–positron colliders is that they typically produce a limited number of particles in the final state, thus facilitating the interpretation of the outcome of the collisions. On the other hand, the total available energy is fixed at the start of the experiment, and it is generally difficult to increase because this would require improving the whole accelerator set-up. Furthermore, electrons and positrons, when accelerated, tend to massively lose energy due to electromagnetic radiation, so that pushing electron–positron machines towards high energies with current accelerator facilities is particularly challenging. Therefore, e^+e^- collisions are not ideal for discovering new particles whose masses are unknown, but are instead useful for precisely measuring the properties of newly discovered particles. This was the case of the LEP machine in its first run (LEP1), operating at $\sqrt{s} = 91.2$ GeV, the mass of the Z boson. LEP1 was focused on the study of the properties of this particle.

Hadron–hadron collisions are mainly aimed at the discovery of new particles. In fact, at high energies, hadrons break apart, and their constituent quarks and gluons undergo elementary highly energetic collisions, producing all sorts of particles. Each parton involved in the collision carries an unknown fraction of the parent hadron's energy, so that the total energy available in the collision is unknown. This property makes it possible to span a continuous range of energies up to the centre-of-mass energy of the hadron–hadron collision, without changing the experimental set-up as in e^+e^- machines. Furthermore, hadrons like protons or antiprotons lose less energy than electrons and positrons through electromagnetic radiation, and hence can be more effectively accelerated to higher energies. Examples of high-energy hadron colliders are SPS [27], the Tevatron [28] and the LHC [29]. The SPS, located at CERN, at its time of operation was a proton–antiproton collider, and is most famous for the discovery of the W [30, 31] and Z [32, 33] bosons by the two experiments UA1 [34] and UA2 [35]. The Tevatron, at Fermilab, which terminated its operations in 2011, used proton and antiproton beams, with centre-of-mass energy $\sqrt{s} = 1.8$ TeV in its first run and $\sqrt{s} = 1.96$ TeV after an upgrade. The machine hosted the two experiments CDF [36] and D0 [37]. Among its most important results is the discovery of the top quark [38, 39]. The LHC is a proton–proton collider located at CERN. It first ran at $\sqrt{s} = 7$ TeV and $\sqrt{s} = 8$ TeV, and has been recently upgraded to reach the centre-of-mass energy of 13.6 TeV. In its first run, the LHC discovered a spin-0 particle whose properties are compatible with the Higgs boson of the Standard Model [40, 40]. The LHC, in particular through the experiments ATLAS [41] and CMS [42], is the machine which, at the moment, is expected to discover new physics beyond the Standard Model.

The different characteristics of electron–positron and hadron–hadron collisions have consequences on the kinematic variables that are typically used in physics analyses. Before discussing these differences, it is useful to quickly review the different parts of a high-energy physics detector. Close to the collision point there is a tracker, able to precisely determine the direction of charged particles. This makes it possible to measure charged particle three-momenta by bending their trajectories with a magnetic field. After the tracker there are two detectors called calorimeters devoted to the measurement of particle energies. The first is the so-called 'electromagnetic' calorimeter, where photons and electrons lose all their energy. Hadrons instead lose only part of their energy inside the electromagnetic calorimeter, so that their energy determination requires an additional detector, called the 'hadronic' calorimeter, where all hadrons are supposed to stop. Muons are the only charged particles that escape the hadronic calorimeter. Their three-momenta are measured through muon detectors, which represent the outermost part of a high-energy detector. Neutrinos are not detected at all, and contribute to the so-called missing energy.

Typically, in high-energy electron–positron colliders, the reference frame of the laboratory coincides with the centre-of-mass frame of the collision. Therefore, one naturally stores the energy and the three-momentum of each particle in that reference frame. In hadron collisions, beam particles break apart, and the energy of each elementary collision is not known. It is therefore very important to use kinematic variables that transform as simply as possible under Lorentz boosts in the beam direction (which sets for us the z-direction). One of these quantities is the transverse momentum of each particle with respect to the beam, which is invariant with respect to such boosts. For a particle of momentum $p = (E, p_x, p_y, p_z)$, its transverse momentum is identified by its magnitude $p_t \equiv \sqrt{p_x^2 + p_y^2}$ and its azimuthal angle $\phi = \arctan(p_y/p_x)$. Another useful variable is the rapidity y, defined, given a momentum p, as

$$y \equiv \frac{1}{2} \ln \frac{E + p_z}{E - p_z}. \tag{1.6}$$

If a particle is massless, its rapidity is related to the angle θ that the particle three-momentum forms with the beam axis, as follows

$$y = \frac{1}{2} \ln \frac{1 + \cos\theta}{1 - \cos\theta} = -\ln \tan\frac{\theta}{2}. \tag{1.7}$$

The angular variable $-\ln\tan(\theta/2)$ is called pseudorapidity and denoted by η. For massless particles, rapidity and pseudorapidity coincide. Therefore, a particle close to the beam setting the positive (negative) z-direction has a large positive (negative) rapidity. Zero rapidity corresponds to a particle whose three-momentum is only transverse to the beam. Under a boost along the beam direction, the rapidity of each particle shifts by a constant quantity, so that differences in rapidity are invariant under longitudinal boosts. Therefore, in hadron collisions, it is natural to use transverse momentum and rapidity as kinematic variables for each particle. However, since detectors are typically sensitive to particle directions and energy deposits, it is also customary to give information about each particle's pseudorapidity η and transverse energy $E_t = E \sin\theta$, with E the particle's energy. Obviously, the transverse energy of a massless particle is the magnitude of its transverse momentum. An example of how to describe an event in terms of the aforementioned kinematic variables can be seen in the bottom right-hand corner of figure 1.1. There, the horizontal plane corresponds to the pseudorapidity–azimuth plane (η–ϕ). Each square in that plane corresponds to a hadronic calorimeter cell (one of the segments in which the hadronic calorimeter is divided). The vertical axis instead represents the transverse-energy deposit in each cell. All the energy deposits within each of the coloured circles in the η–ϕ plane are considered to build up a jet.

Another difference between e^+e^- and hadron–hadron collisions, which is particularly relevant for jet physics, is which hadrons can actually be observed. In e^+e^- colliders, only a negligible fraction of hadrons can fall in the tiny angular region around the beam pipe which is not covered by detectors. Therefore, one can reasonably assume that all hadrons are observed in electron–positron colliders. This is in contrast with high-energy hadronic collisions, where the colliding particles are coloured quarks and gluons. QCD radiation from incoming partons is very collimated around the beam direction, which hence contains many hadrons with an interesting dynamics. In hadron–hadron collisions, it is therefore important to consider the actual rapidity range spanned by the various parts of a detector. In fact, the tracker is usually placed in a central angular region, for instance $|\eta| \lesssim 1.5$ at the Tevatron and $|\eta| \lesssim 2.5$ at the LHC. The hadronic calorimeter extends further, for instance up to $|\eta| \simeq 3$ at the Tevatron and $|\eta| \simeq 4.5$ at the LHC. The cells of the hadronic calorimeter do not give access to the same angular resolution as the central tracker, especially in the most forward and backward regions, where hadrons from beam fragmentation are most likely to fall. This means that in hadron collisions, the basic objects that are measured are not individual particles, but rather pseudoparticles, reconstructed out of the energy deposited in the cells of electromagnetic and hadronic calorimeters. It is then natural to try to construct objects that are independent of the fine details of the detectors. Hadronic jets offer a viable solution to this problem, which is yet another reason why they play such an important role in high-energy physics.

References

[1] https://twiki.cern.ch/twiki/bin/view/AtlasPublic/EventDisplayStandAlone

[2] Workman R L *et al* (Particle Data Group Collaboration) 2022 Review of particle physics *Prog. Theor. Exp. Phys.* **2022** 083C01

[3] Breidenbach M, Friedman J I, Kendall H W, Bloom E D, Coward D H, DeStaebler H C, Drees J, Mo L W and Taylor R E 1969 Observed behavior ofhighly inelastic electron-proton scattering *Phys. Rev. Lett.* **23** 935–9

[4] Feynman R P 1969 Very high-energy collisions of hadrons *Phys. Rev. Lett.* **23** 1415–7

[5] Gell-Mann M 1964 A schematic model of baryons and mesons *Phys. Lett.* **8** 214–5

[6] Gross D J and Wilczek F 1973 Ultraviolet behavior of nonabelian gauge theories *Phys. Rev. Lett.* **30** 1343–6

[7] Politzer H D 1973 Reliable perturbative results for strong interactions? *Phys. Rev. Lett.* **30** 1346–9

[8] Hanson G *et al* 1975 Evidence for jet structure in hadron production by $e^+ e^-$ annihilation *Phys. Rev. Lett.* **35** 1609–12

[9] Brandelik R *et al* (TASSO Collaboration) 1979 Evidence for planar events in $e^+ e^-$ annihilation at high-energies *Phys. Lett.* B **86** 243–9

[10] Barber D P *et al* 1979 Discovery of three jet events and a test of quantum chromodynamics at PETRA energies *Phys. Rev. Lett.* **43** 830

[11] Berger C *et al* (PLUTO Collaboration) 1979 Evidence for gluon bremsstrahlung in $e^+ e^-$ annihilations at high-energies *Phys. Lett.* B **86** 418–25

[12] Bartel W *et al* (JADE Collaboration) 1980 Observation of planar three jet events in $e^+ e^-$ annihilation and evidence for gluon bremsstrahlung *Phys. Lett.* B **91** 142–7

[13] Soding P 2010 On the discovery of the gluon *Eur. Phys. J.* H **35** 3–28

[14] Akesson T (Axial Field Spectrometer Collaboration) 1982 Direct evidence for the emergence of jets in events triggered on large transverse energy in *pp* collisions at $\sqrt{s}$ = 63-GeV *Phys. Lett.* B **118** 185–92

[15] Arnison G *et al* (UA1 Collaboration) 1983 Observation of jets in high transverse energy events at the CERN proton–anti-proton collider *Phys. Lett.* B **123** 115–22

[16] Banner M *et al* (UA2 Collaboration) 1982 Observation of very large transverse momentum jets at the CERN anti-p p collider *Phys. Lett.* B **118** 203–10

[17] Salam G P 2010 Towards jetography *Eur. Phys. J.* C **67** 637–86

[18] Khachatryan V *et al* (CMS Collaboration) 2014 Observation of the diphoton decay of the Higgs boson and measurement of its properties *Eur. Phys. J.* C **74** 3076

[19] Aad G *et al* (ATLAS Collaboration) 2013 Measurements of Higgs boson production and couplings in diboson final states with the ATLAS detector at the LHC *Phys. Lett.* B **726** 88–119 [Erratum: 2014 *Phys. Lett.* B **734** 406]

[20] https://twiki.cern.ch/twiki/bin/view/CMSPublic/PhysicsResultsCombined

[21] Naroska B 1987 $E^+ e^-$ physics with the JADE detector at PETRA *Phys. Rep.* **148** 67

[22] Schachter M-J *et al* (CELLO Collaboration) 1981 CELLO: a new detector at Petra *Phys. Scr.* **23** 610–22

[23] https://aleph.web.cern.ch/aleph/

[24] https://delphi-http://www.web.cern.ch/delphi-www/

[25] https://l3.web.cern.ch/l3/

[26] http://opal.web.cern.ch/Opal/

[27] https://home.cern/science/accelerators/super-proton-synchrotron

[28] https://www.fnal.gov/pub/tevatron/tevatron-accelerator.html

[29] https://home.cern/science/accelerators/large-hadron-collider

[30] Arnison G *et al* (UA1 Collaboration) 1983 Experimental observation of isolated large transverse energy electrons with associated missing energy at $\sqrt{s} = 540$ GeV *Phys. Lett.* B **122** 103–16

[31] Banner M *et al* (UA2 Collaboration) 1983 Observation of single isolated electrons of high transverse momentum in events with missing transverse energy at the CERN anti-p p collider *Phys. Lett.* B **122** 476–85

[32] Arnison G *et al* (UA1 Collaboration) 1983 Experimental observation of lepton pairs of invariant mass around 95-GeV/c**2 at the CERN SPS collider *Phys. Lett.* B **126** 398–410

[33] Bagnaia P *et al* (UA2 Collaboration) 1983 Evidence for $Z^0 \to e^+e^-$ at the CERN $\bar{p}p$ collider *Phys. Lett.* B **129** 130–40

[34] https://home.cern/science/experiments/ua1

[35] https://home.cern/science/experiments/ua2

[36] https://www.fnal.gov/pub/tevatron/experiments/cdf.html

[37] https://www.fnal.gov/pub/tevatron/experiments/dzero.html

[38] Abe F *et al* (CDF Collaboration) 1995 Observation of top quark production in $\bar{p}p$ collisions *Phys. Rev. Lett.* **74** 2626–31

[39] Abachi S *et al* (D0 Collaboration) 1995 Observation of the top quark *Phys. Rev. Lett.* **74** 2632–7

[40] Aad G (ATLAS Collaboration) 2012 Observation of a new particle in the search for the Standard Model Higgs boson with the ATLAS detector at the LHC *Phys. Lett.* B **716** 1–29

[41] Chatrchyan S *et al* (CMS Collaboration) 2012 Observation of a New boson at a mass of 125GeV with the CMS experiment at the LHC *Phys. Lett.* B **716** 30–61

[42] https://atlas.cern/

[43] https://cms.cern/detector

Chapter 2

Jet algorithms

Let us consider an event like the one displayed in the left-hand panel of figure 2.1, in which we recognise the presence of hadronic jets, and ask ourselves how many jets we observe. One might say it is clearly two, but for instance the (pale blue) tracks on the left-hand side of the picture could be considered to form a jet on their own. In fact, this event is classified by ALEPH as a four-jet event, but on what basis? If we consider the event as containing two jets, which hadrons have to be included in each jet? If answering such questions might seem an easy task for the event in the left-hand panel of figure 2.1, what about the one in the right-hand panel, which contains hadrons spread all over the detectors? Even if we have succeeded in assigning each hadron to a jet in the events shown in figure 2.1, we have to repeat the same procedure for every event. We definitely need a set of rules to establish how many jets each event contains, and which hadrons have to be assigned to each jet. Such a set of rules is called a 'jet algorithm'. There exist many jet algorithms, and choosing one or the other depends crucially on the kind of information that we wish to extract from a set of events. Before delving into the details of the various jet algorithms, let us focus on what a jet algorithm should intuitively do. Jet events arise from the production of highly energetic (hard) quarks and gluons, which later on transform into collimated bunches of hadrons, through a mechanism that will be explained in the next chapter. It is natural to require that the number of jets we observe, as well as their energy–momentum flow, reflect the properties of the hard quarks and gluons that were initially produced in the elementary collision. For instance, if an event originates from the production of a hard quark–antiquark pair in electron–positron annihilation, a good jet algorithm should produce as output two jets, with momenta very close to those of the quark and antiquark that initiated the event.

Sterman–Weinberg jets. An example on how to map final-state hadrons into jets was provided by Sterman and Weinberg [2], who developed essentially the very first jet algorithm. The algorithm, devised for e^+e^- annihilation, works as follows: an

doi:10.1088/978-0-7503-4737-2ch2

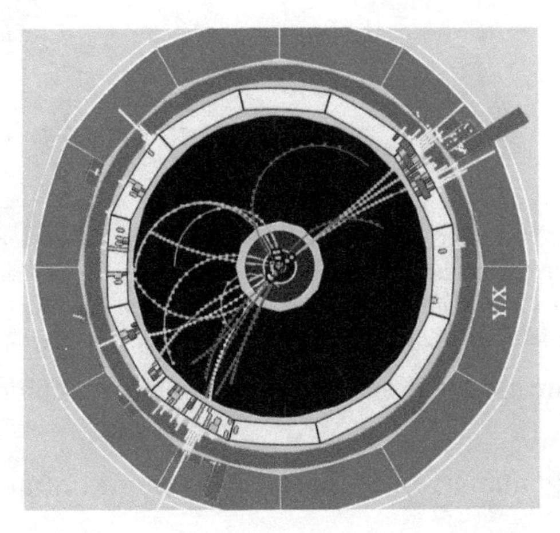

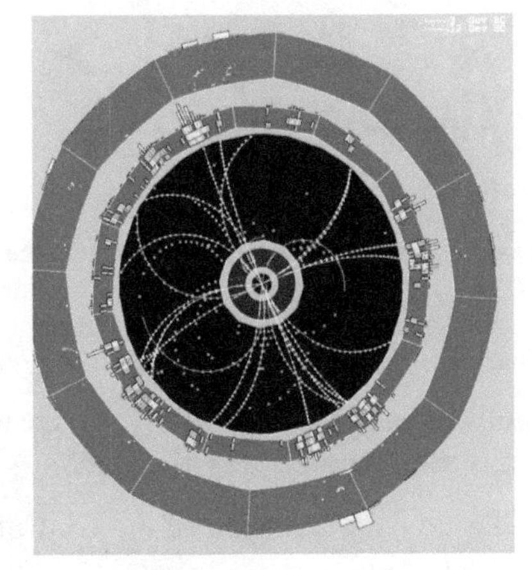

Figure 2.1. Two four-jet events from the event display of the ALEPH collaboration at LEP [1]. Copyright © ALEPH.

event is classified as having two jets if all but at most a fraction ϵ of the total energy of produced hadrons is contained inside two cones of opening angle δ. If ϵ is sufficiently small, the two cones will contain the quark and antiquark produced in the hard collision. This simple example already highlights an important aspect of jet algorithms: they depend on parameters, varying which one changes the fraction of hadrons that get included in each jet. Another important feature of Sterman–Weinberg jets is that, given ϵ and δ, the fraction of events having two jets, the so-called two-jet rate, is a well-defined observable, and can be computed using relativistic quantum mechanics within the framework of quantum chromodynamics (QCD), the theory of quarks and gluons. What is the property of Sterman–Weinberg jets that makes this possible? To answer this question we need to consider jets immediately before quarks and gluons transform into hadrons. In QCD, the probability of emitting a gluon that has exactly zero energy is infinite. This pathological behaviour is referred to as 'soft' divergence. Similarly, one obtains an infinite result if one parton (quark or gluon) splits into two (or more) parallel (collinear) partons, giving rise to a 'collinear' divergence. Fortunately, quantum fluctuations (virtual corrections), for instance the fact that a gluon is emitted and reabsorbed before it is observed, lead to the same kind of divergences, but with opposite sign. As a consequence, if a physical observable is affected in the same way by an infinitely soft gluon and the corresponding quantum fluctuation, the infinities will cancel and a QCD calculation in terms of quarks and gluons will give a finite result. If an observable is to be calculable in QCD, a similar cancellation should occur for collinear divergences as well. For instance, in the case of Sterman–Weinberg jets, for a given value of ϵ and δ, after the addition of a zero-energy gluon the event will still be considered a two-jet event. Similarly, a quantum fluctuation cannot change the number of jets. Therefore, when computing the two-jet rate, infinities cancel between real (gluon emission) and virtual (quantum) corrections. This is illustrated in figure 2.2.

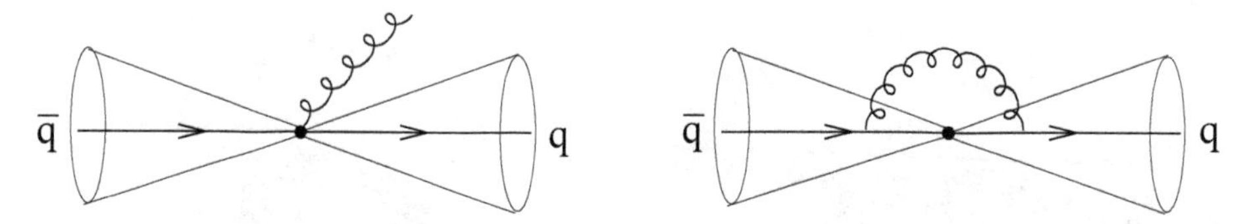

Figure 2.2. Representative diagrams that contribute to the rate of Sterman–Weinberg jets at the first order in QCD perturbation theory. The emission of an infinitely soft gluon (left) does not change the amount of energy outside the two cones around the primary quark–antiquark pair. The same happens in the case of a quantum fluctuation (right). This ensures the cancellation of infrared singularities between the two kinds of diagrams.

Infrared and collinear safety. To ensure cancellation of soft and collinear infinities, jet momenta should stay the same

1. after the addition of an arbitrary number of infinitely soft partons (infrared safety);
2. after an arbitrary number of collinear splittings (collinear safety).

Jet algorithms that satisfy both requirements are called infrared and collinear (IRC) safe algorithm. Sterman–Weinberg jets are IRC safe.

IRC safety is the property that ensures that jets defined at detector level (for instance using calorimetric cells as inputs), at hadron level (using hadrons with a lifetime up to an agreed value), and at parton level, (obtained from quarks and gluons), are essentially the same. In fact, it is possible to show that, for IRC safe observables, the reshuffling of momenta due to hadronisation leads to effects that are suppressed by inverse powers of the typical momentum scale of the process under consideration, for instance the transverse momentum of a given jet. The higher this scale, the closer hadronic IRC safe observables will be to the corresponding partonic ones.

IRC safe jet algorithms are also extremely useful from an experimental perspective. In fact, a huge issue in implementing jet algorithms experimentally is the fact that calorimetric cells, especially at hadron colliders and in the forward and backward regions, cannot resolve the energy deposit of single particles. Therefore, jets can only be defined using the transverse energy, the pseudo-rapidity and the azimuth of individual calorimetric cells as inputs, rather than momenta of individual hadrons. However, for IRC safe jet algorithms, a soft gluon will give an energy deposit which will not alter the momenta of the existing jets. Similarly, collinear splittings, giving energy deposits in the same calorimetric cell, will be clustered inside the same jet. In practice, this implies that the IRC safe jet definitions will be independent of the details of the detector, up to corrections that vanish as a power of the detector resolution. The granularity of calorimeters represents such an issue that, in hadron collisions, jets are the main objects that enter physics analyses. To be as insensitive as possible to detector effects, jet algorithms must then be IRC safe.

Good jet algorithms. In the early 1990s, a document known as the 'Snowmass Accord' [3] summarised the desired features of jet algorithms. Specifically, a good jet algorithm should:

1. be simple to implement in an experimental analysis;
2. be simple to implement in theoretical calculations;

3. be defined at any order of perturbation theory;
4. yield finite cross sections at any order of perturbation theory;
5. yield a cross section that is relatively insensitive to hadronisation.

IRC safety automatically ensures that the last three conditions are satisfied.

Concerning point 2 above, due to the fact that most QCD calculations are performed numerically via Monte Carlo procedures that simulate collider events, a jet clustering algorithm can be arbitrarily complicated. However, in order to have an understanding of the properties of the algorithm, it might be useful if jet observables (e.g. jet rates) could to some extent be computed analytically. This is why procedures understandable by humans are in general preferred to other procedures.

Point 1 deserves special attention. As for point 2, there is no conceptual problem in letting a jet algorithm process a set of particle momenta. However, in environments with lots of particles, such as high-luminosity hadron colliders, or even heavy-ion colliders, the speed of an algorithm can become a real issue. This is why a lot of effort has been devoted to designing algorithms that are not only IRC safe, but whose computational complexity does not grow too fast with the number of input particles. Another important experimental issue is the elimination of background. This can arise due to particles that do not belong to the event under consideration, for instance originating from secondary collisions occurring close in time with respect to a primary interesting collision (the so-called 'pile-up' (PU)), or simply from detector noise. Such backgrounds are easier to eliminate if jets have a fixed shape, for instance if each jet is enclosed in a circle in the η–ϕ plane. Achieving this last property is a highly non-trivial task, as will become clear at the end of section 2.1.1.

Given this general overview, we will now discuss different jet algorithms. In section 2.1, we will present the two main families of algorithms currently in use, namely cone and sequential algorithms. Rather than giving a historical overview, we will concentrate on the distinctive features of both families. In section 2.2 we will discuss some common experimental issues that one has to confront when working with jets, namely the problems of jet energy scale and the elimination of a uniform noise. We will conclude in section 2.3 presenting a number of more recent ideas on jet algorithms.

2.1 Cone or sequential algorithms?

In this section we describe the two main classes of algorithm used at colliders, namely cone and sequential algorithms.

2.1.1 Cone algorithms

When looking at events containing jets, it comes natural to draw cones around the most energetic deposits in detectors and identify a jet as the set of particles within one of those cones. Besides this intuitive pictorial representation, how does one practically construct those cones with an algorithmic procedure? Here we focus on hadron colliders, where cone algorithms have been widely applied for many years.

Figure 2.3. The display of an event recorded by the ATLAS detector at CERN. The bottom right corner shows the lego plot of the event [4]. ATLAS Experiment, Copyright 2014 CERN.

There, a natural way of looking at events is the two-dimensional space defined by pseudo-rapidity and azimuth (the η–ϕ plane), where particles appear as spots, having activated a number of calorimetric cells. Figure 2.3 contains, in the bottom right corner, a so-called 'lego' plot, in which not only is it possible to see the active calorimetric cells in the η–ϕ plane, but also the corresponding transverse energy deposit, represented by the height of the tower above each cell. The basic idea behind cone algorithms at hadron colliders is that one can draw circles in the η–ϕ plane, and the particles whose transverse energy deposits fall inside any of these circles constitute a jet. In three dimensions, the resulting jets will look like cones, hence the name of this family of jet algorithms. The main problem with cone algorithms is that drawing those circles in such a way that the resulting jets are IRC safe is a highly non-trivial task, as will be clear in the following.

Seeded cones. An intuitive way of finding jets is to start by drawing a circle with a fixed radius in the rapidity–azimuth (y–ϕ) plane[1] around the most energetic spot in an event, and consider all the particles inside that circle to build up a jet. More precisely, one considers a list of pseudo-particles, which can be individual particles, calorimetric cells, or even jets resulting from some earlier clustering procedure. One then takes the pseudo-particle with the largest transverse momentum (or transverse energy), and draws around it a circle of radius R, the jet radius, in the y–ϕ plane. All pseudo-particles within that circle are considered to build up a jet. These pseudo-

[1] In the rest of the chapter we will use transverse momenta and rapidity to describe particles in hadron collisions, since these quantities have simple transformation rules under Lorentz boosts. All the quantities that we will introduce can be redefined in terms of transverse energy and pseudo-rapidity, without changing any of the conclusions we will draw.

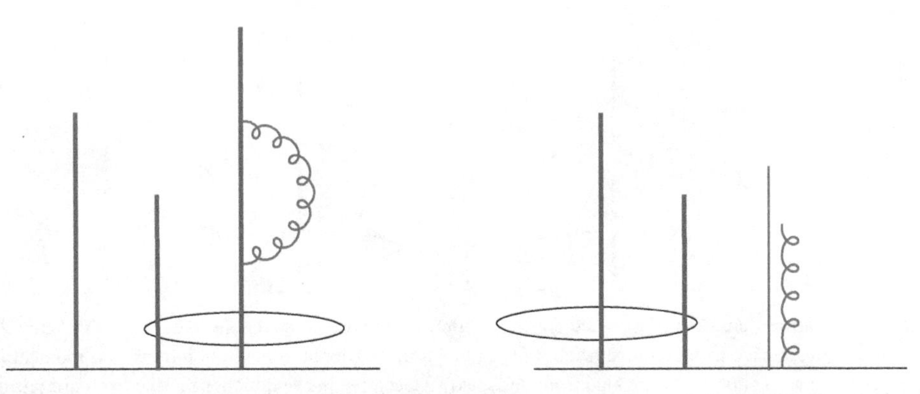

Figure 2.4. Collinear unsafe jet algorithm. Each line represents a parton, and the height of each line is proportional to the parton transverse energy. The horizontal axis can be thought of as the rapidity axis at a fixed azimuth, or vice versa. With a virtual correction (left), the parton on the right is the most energetic, so that a jet is formed by this parton and the central one. If a collinear splitting occurs (right), the parton on the left becomes the most energetic, and will then form a jet with the central parton.

particles are then removed from the original list, and the procedure is repeated until all pseudo-particles have been assigned to a jet. This procedure, known as the 'fixed-cone' algorithm, is the first example of a 'seeded' cone algorithm, which draws circles starting from reference points, or 'seeds', in the y–ϕ plane. This simple example highlights already one of the main problems of seeded cone algorithms, which is how to draw cones so as to obtain an IRC safe procedure. The cone algorithm we have just described, similar to the ones used by the UA1 and UA2 experiments at the SPS [5], is unfortunately collinear unsafe. In fact, suppose that the particle with the largest transverse momentum splits into two particles, neither of which is the one with the highest transverse momentum. This splitting changes the jet clustering sequence. As a consequence, the momenta of the resulting jets will be different from the situation where no splitting has occurred, for instance in the presence of quantum fluctuations. In this case, real and virtual corrections can give rise to different jets, and infinities will not cancel between them. An example of configurations that lead to a non-cancellations of collinear singularities is illustrated pictorially in figure 2.4.

Before discussing attempts to refine the jet-finding procedure, we remark that the quantity that should not change in an IRC safe cone algorithm in hadron collisions is not the total number of jets, but rather the jets that have a transverse momentum above a given threshold, which we call the 'hard' jets. In fact, nothing can prevent an infinitely soft gluon sufficiently far away from all the other jets to give rise to an infinitely soft jet. This however does not change the number of hard jets, neither their momenta.

A further attempt to construct an IRC safe cone algorithm uses the momenta of all pseudo-particles as seeds [6]. One starts with any pseudo-particle and constructs a jet axis out of the momenta of all pseudo-particles within a distance R from it in the y–ϕ plane, for instance as the direction of the sum of the momenta of the selected pseudo-particles. The resulting axis is a new trial direction, and the procedure is

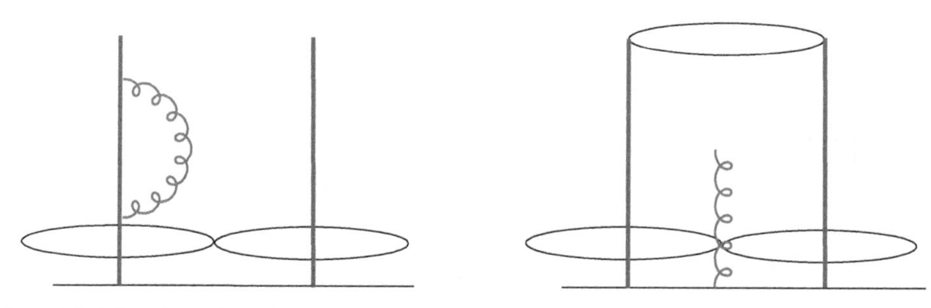

Figure 2.5. Infrared unsafety of a seeded cone algorithm using particle directions as seeds. As in figure 2.4, each line represents a parton, and the height of each line is proportional to the parton transverse momentum. With a virtual correction (left), the algorithm finds two jets, centred in the directions of the two hard partons (the long blue lines). With an infinitely soft gluon (the red curly line) in the middle of the two hard partons (right), the algorithm finds another jet that includes both hard partons.

repeated until the pseudo-particle content of the set does not change any more. At this point, another pseudo-particle is used as seed, until all trial directions have been used. Such algorithms are commonly referred to as 'iterative-cone' finders. Unfortunately, the procedure we have just described is not infrared safe. In fact, if we have two pseudo-particles with similar transverse momentum, separated by a distance that is less than twice the jet radius R, the algorithm will find two cones, each centred around the two pseudo-particles. However, when a soft gluon is emitted between the two energetic pseudo-particles, the algorithm will find another cone which will include the two energetic pseudo-particles. The algorithm is then infrared unsafe, because quantum corrections cancelling the infinity coming from soft gluon emission do not give rise to new seeds to find the hard jets [7] (see figure 2.5). An attempt to fix this problem could be to add more seeds, for instance considering also the midpoint between any pair of pseudo-particles as an additional trial direction [8]. Unfortunately, as explained for instance in [9], even for this choice of seeds it is always possible to find configurations in which the addition of an infinitely soft gluon leads to finding a new hard jet. At the moment there exists no IRC safe seeded cone algorithms, although a no-go theorem stating that it is impossible to have an IRC safe cone algorithm with a *finite* number of seeds has never been openly formulated.

Stable cones. A procedure that provides IRC safe cone algorithms involves the concept of a stable cone of radius R, which is the set of all pseudo-particles p_i, which are within a distance R in the y–ϕ plane from a jet axis (y_J–ϕ_J):

$$(y_i - y_J)^2 + (\phi_i - \phi_J)^2 < R^2, \tag{2.1}$$

with the jet axis constructed out of the momenta of the particles inside the stable cone. In practice, a stable cone is a circle in the y–ϕ plane with its centre coinciding with the jet axis.

Neither a collinear splitting, nor a soft emission can change the momentum of a stable cone. Therefore, if we identify each stable cone with a jet, the number and momenta of the jets with transverse momentum above a given threshold is an IRC safe quantity. A way to look for all stable cones is to consider all possible circles of radius R that one can draw, and for each one check if the position of the corresponding jet axis in the y–ϕ plane coincides with the centre of the circle: then we have found a stable cone. In practice this is not feasible, implying that more efficient procedures have to be devised.

The problem of finding all stable cones has been solved in a general way using seedless algorithms. From an experimental point of view, one can draw circles centred in each cell of the hadronic calorimeter, and check whether each corresponds to a stable cone [8]. This is the closest equivalent to a seedless algorithm that draws all possible circles, but is quite expensive from a computational point of view, because it requires $\mathcal{O}(N_{\text{cells}}n)$ steps, where N_{cells} is the number of calorimetric cells, and n the typical number of particles within a cone. If one has information on the momenta of all pseudo-particles, one can consider all possible subsets of pseudo-particles and check whether each subset gives a stable cone [10]. In this case it is obvious that all stable cones will be found. However, the number of subsets that can be formed out of N pseudo-particles is 2^N, so that the algorithm becomes computationally impractical for a large number of particles. A much faster procedure to find stable cones is provided by the SISCone (Seedless Infrared Safe Cone) algorithm [9]. The main idea behind SISCone is to exploit methods borrowed from computational geometry to efficiently move circles of radius R around the y–ϕ plane until all stable cones are found. In fact, this can be performed in $\mathcal{O}(Nn \ln n)$ steps, where again N is the total number of pseudo-particles, and n the typical number of particles within each cone. Furthermore, it is possible to show that all hard stable cones found with the SISCone procedure are IRC safe.

Overlapping cones. After all the stable cones have been found, many of them will have particles in common, i.e. they will overlap. One then needs a procedure to decide how to move from a set of stable cones, which at this stage are commonly referred to as 'proto-jets', to the list of the final jets. The current way to deal with overlapping cones is the split-merge procedure [8]. One starts with the proto-jet for which the scalar sum of the transverse momenta of its constituents is the largest. Let us call this proto-jet p_a, and look for the closest (in the y–ϕ plane) proto-jet p_b that overlaps with p_a. If this is not found, p_a is considered to be a jet, and removed from the list of proto-jets. Otherwise, the two proto-jets are merged in a single proto-jet if the scalar sum of transverse momenta of the shared particles is more than a fraction f (normally chosen to be 0.5 of 0.75) of the scalar sum of transverse momenta of proto-jet p_b. If this is not the case, the shared particles are assigned to either of the proto-jets, currently each particle to the jet whose axis is closer. This is repeated until the selected proto-jet has no overlap with any other proto-jet, in which case it is called a jet and removed from the list of proto-jets. The split-merge procedure continues with the other proto-jets until no proto-jets are left. A viable alternative is the split-drop procedure [11], in which the shared particles are attributed to the proto-jet with the largest scalar sum of constituents' transverse momenta, and the remaining particles belonging to the jets with smaller scalar sum of constituents' transverse momenta are just eliminated. This is an example of a procedure in which 'dark towers' are created, i.e. objects that are not clustered with any jet. Dark towers

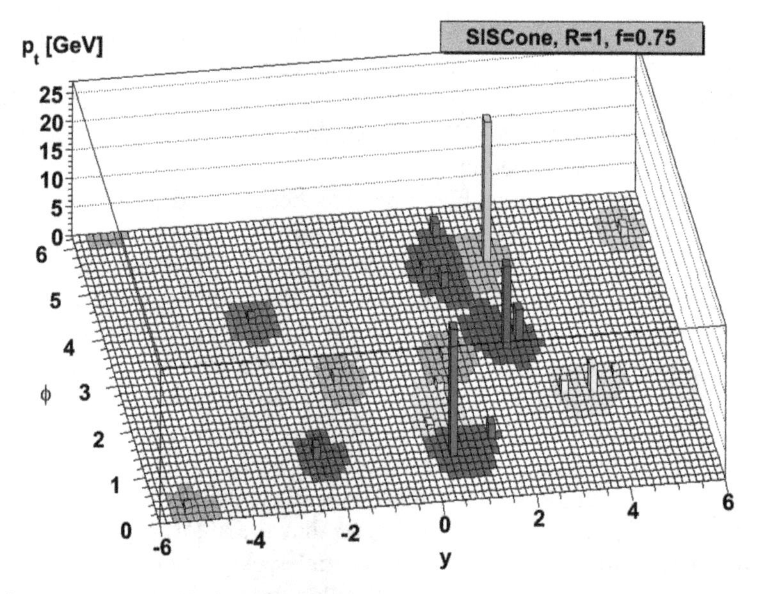

Figure 2.6. Jets reconstructed with the SISCone algorithm, with radius $R = 1$ and a split-merge procedure to deal with overlapping cones corresponding to $f = 0.75$. Cells with the same colour are clustered within the same jet. Reproduced with permission from [14].

are common in cone algorithms, and have been dealt with in a number of different ways [12]. The simplest seems to be to run the jet algorithm repeatedly over the pseudo-particles that have been eliminated through the split-drop procedures, until all dark towers have been assigned to a jet [13].

An unwanted feature of the split-merge procedure is that, in the presence of many soft particles, the shapes of well-separated hard cones are not perfect circles in the y–ϕ plane, as shown in figure 2.6. Not only is their shape irregular, but it is also known only after the whole jet-finding procedure is terminated. This makes it difficult to subtract a uniform background noise on an event-by-event basis, as will be discussed in more detail in section 2.2. Surprisingly, circular cones can instead be achieved using sequential algorithms, the topic of the next subsection.

2.1.2 Sequential algorithms

Sequential algorithms reconstruct jets by clustering particles pairwise until no particles are left. We will first discuss the general features of sequential algorithms developed for e^+e^- collisions, and later on introduce their counterparts in hadron collisions. As for cone algorithms, the starting point is a set of pseudo-particles, which can be either true particles, or the result of the recombination of one or more particles.

Electron–positron annihilation. In the simplest version of e^+e^- sequential algorithms, one considers all pairs of pseudo-particles and finds p_i and p_j, the ones for which an IRC safe distance measure y_{ij} is the smallest. One then introduces a resolution y_{cut}, and if $y_{ij} < y_{\text{cut}}$, the two pseudo-particles are recombined into a new

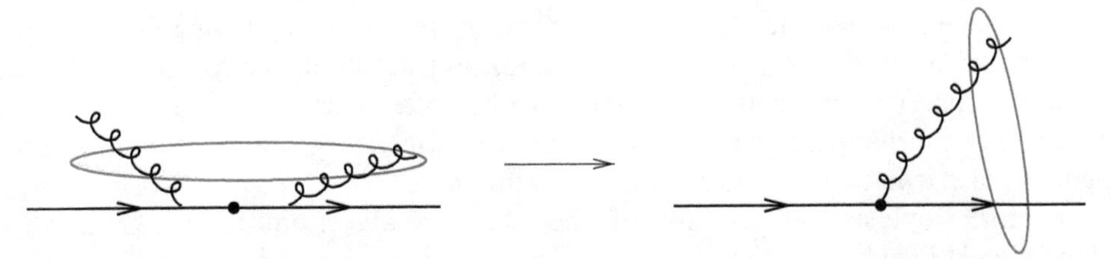

Figure 2.7. Creation of a soft large-angle jet out of two gluons with the JADE algorithm. The red hoops indicate which partons will be clustered together by the jet algorithm. The gluon on the left does not get clustered with the parton to which it is collinear.

pseudo-particle, for instance by just adding their four-momenta p_i and p_j. The procedure is then repeated with the remaining pseudo-particles. If, at any step, all y_{ij} distances are larger than y_{cut}, the algorithm terminates, and the jets are the pseudo-particles left at this stage. The physical observable that is generally measured is the n-jet rate $R_n(y_{\text{cut}})$, defined as the fraction of events for which, for a given value of y_{cut}, the procedure returns n jets.

A simple example of an IRC safe distance measure is the invariant mass of p_i and p_j. This is implemented in the JADE algorithm [15]:

$$y_{ij}^{(\mathrm{J})} \equiv \frac{(p_i + p_j)^2}{Q^2}, \tag{2.2}$$

where Q is the centre-of-mass energy of the e^+e^- collision. This measure is by construction IRC safe, because it vanishes when either p_i or p_j is soft, or when the pair is collinear. Therefore, we can use the measure $y_{ij}^{(\mathrm{J})}$ to construct a sequential jet algorithm. Unfortunately, the JADE algorithm has an unwanted feature, which is easily understood if one considers configurations where a hard quark–antiquark pair, flying in opposite directions, is accompanied by two soft gluons, one collinear to the quark, and the other to the antiquark. In this situation, depicted in figure 2.7, it is possible that the algorithm, instead of clustering each gluon to the parton to which it is collinear, clusters the two gluons together, creating a soft large-angle pseudo-particle, which later on will be clustered either with the quark or with the antiquark. Therefore, one of the two gluons will be attracted towards a particle which is far away in angle. This is not ideal, because a jet initiated by a hard parton should include as much as possible the partons successively radiated from it, and should not attract radiation from other hard partons. This feature creates also complications from the point of view of all-order QCD calculations, as will be discussed in the next chapter.

An improved distance that provides a solution to the problem of the JADE algorithm is provided by the Durham algorithm [16]:

$$y_{ij}^{(\mathrm{D})} = 2\frac{\min(E_i^2, E_j^2)}{Q^2}(1 - \cos\theta_{ij}), \tag{2.3}$$

with E_i and E_j the energies of pseudo-particles p_i and p_j and θ_{ij} their relative angle. The above distance is, as needed, IRC safe, and at small angles reduces to the relative transverse momentum (squared) of the softer particle with respect to the harder. The Durham algorithm does not create spurious large-angle jets, so collinear bunches of particles get clustered into the same jet.

A more sophisticated variant of the Durham algorithm is the Cambridge algorithm [17]. The Cambridge clustering procedure is aimed at reconstructing backward the typical sequence of gluon emissions, which occur predominantly at successively decreasing angles. At each step, the Cambridge algorithm finds the pair of pseudo-particles with the smallest *angular* distance $v_{ij} = (1 - \cos\theta_{ij})$. This distance is not infrared safe, so we cannot not simply follow the sequential procedure described at the beginning of this section. Instead, one computes the Durham distance $y_{ij}^{(\mathrm{D})}$ of the selected pair, as given by equation (2.3). If $y_{ij}^{(\mathrm{D})} < y_{\mathrm{cut}}$, the two particles are merged, otherwise the object with the smaller energy is stored as a jet, and removed from the list of pseudo-particles. The procedure stops until no merging can be performed, in which case the event is classified as having as many jets as the number of pseudo-particle left.

Hadron collisions. Both the Durham [18] and the Cambridge [19] algorithms have been generalised to hadron collisions. In this context, the algorithms are usually run in a different way with respect to e^+e^- annihilation, in that, as for cones, one is interested in the number of hard jets in an event, even without having specified a jet resolution [20]. Let us consider the Durham algorithm first, which is called the k_t algorithm in the context of hadron collisions. At each step, one considers the pair of pseudo-particles p_i and p_j with the smallest distance (which we will call the 'k_t distance')

$$d_{ij} = \min(p_{ti}^2, p_{tj}^2)\frac{\Delta R_{ij}^2}{R^2}, \qquad \Delta R_{ij}^2 = (y_i - y_j)^2 + (\phi_i - \phi_j)^2, \tag{2.4}$$

where R is a parameter that plays the role of a jet radius[2]. One then finds the distance of p_i and p_j from the beam:

$$d_{i\mathrm{B}} = p_{ti}^2, \qquad d_{j\mathrm{B}} = p_{tj}^2. \tag{2.5}$$

If the minimum among d_{ij}, $d_{i\mathrm{B}}$, $d_{j\mathrm{B}}$ is the mutual distance d_{ij}, then p_i and p_j are recombined into a single pseudo-particle. Otherwise, if that minimum is $d_{i\mathrm{B}}$ (or $d_{j\mathrm{B}}$), pseudo-particle p_i (or p_j) is removed from the list of pseudo-particles and added to the list of jets. Similarly, for the Cambridge algorithm, known as Cambridge/Aachen algorithm in this context [19], the distance d_{ij} is given just by $\Delta R_{ij}^2/R^2$ and $d_{i\mathrm{B}} = d_{j\mathrm{B}} = 1$, and the procedure runs in the same way as for the k_t-algorithm. Notice that, in hadron collisions, the distance measures need to be collinear safe, but do not need to be infrared safe. In fact, as for cone algorithms, the quantity that has

[2] Note that, unlike in e^+e^- annihilation, the distance in equation (2.4) is dimensionful, given the fact that, in hadron collisions, it is not immediate to identify a typical hard scale.

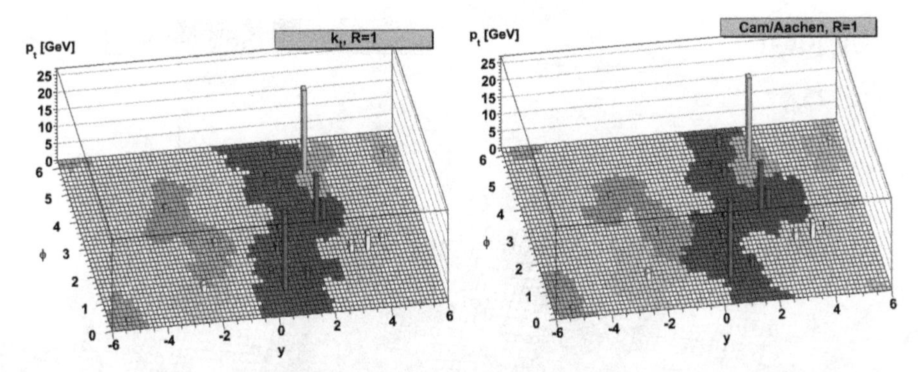

Figure 2.8. Jets reconstructed with the k_t (left) and the Cambridge/Aachen (right) algorithms, corresponding to the same jet radius $R = 1$. Cells with the same colour are clustered within the same jet. Reproduced with permission from [14] © IOP Publishing. All rights reserved.

to be IRC safe is the momenta of the jets that have a transverse momentum above a given threshold.

As can be seen from the above discussion, sequential algorithms are easier to understand from a theoretical point of view, in that they are manifestly IRC safe, and do not present the issue of the same pseudo-particle being assigned to different jets, as happens for overlapping cones. In practice, until recently, sequential algorithms had some practical issues, which made them less attractive than cone algorithms for hadron collisions.

The first problem of sequential algorithms is an experimental one. In the presence of many soft particles, algorithms such as the k_t or the Cambridge/Aachen give rise to jets whose boundary in the y–ϕ plane is very irregular, as can be seen from the examples in figure 2.8. Furthermore, the shape of this boundary is known only *a posteriori*, after the algorithm has finished reconstructing all the jets. This makes it difficult to subtract from each jet the contribution of a uniform background noise. However, this problem is elegantly solved by considering the family of generalised-k_t algorithms identified by the distances

$$d_{ij}^{(p)} = \min\left(p_{ti}^{2p}, p_{tj}^{2p}\right)\frac{\Delta R_{ij}^2}{R^2}, \quad d_{i\text{B}}^{(p)} = p_{ti}^{2p}, \quad d_{j\text{B}}^{(p)} = p_{tj}^{2p}, \tag{2.6}$$

with p a parameter. For $p = 1$ and $p = 0$ one gets back the k_t and Cambridge/Aachen algorithms respectively. For $p = -1$ one obtains a new procedure, called the 'anti-k_t' algorithm. Even in the presence of many soft particles, well-separated hard jets obtained with the anti-k_t algorithm, as shown for instance in figure 2.9, have precisely the shapes of circles in the y–ϕ plane [14]. This unexpected feature explains why all LHC experiments use the anti-k_t as a default choice. An intuitive explanation of this fact resides in that, while the k_t-algorithm starts clustering particles starting from the softest ones, the anti-k_t does the opposite, i.e. it clusters soft particles around the hardest ones, whose direction in the y–ϕ plane is not

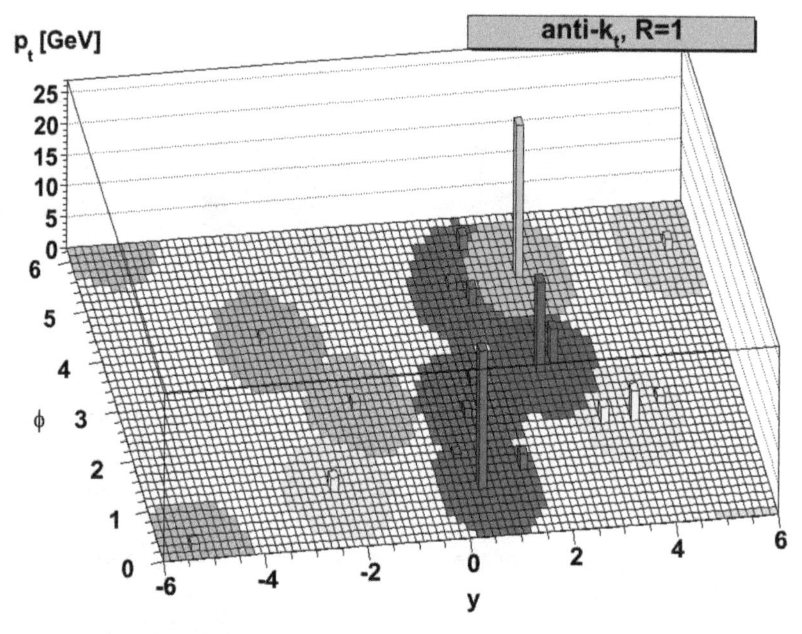

Figure 2.9. Jets reconstructed with the anti-k_t algorithm. As in previous examples, cells with the same colour are clustered in the same jet. Reproduced with permission from [14] © IOP Publishing. All rights reserved.

affected by the emission of soft particles. Note also that there is a fundamental difference between the anti-k_t clustering and the fixed-cone algorithm described in section 2.1.1. In the anti-k_t algorithm, two hard collinear particles are first clustered together before clustering other softer particles around their direction. This is because the distance in equation (2.6) is collinear safe, whereas the order with which seeds are selected in fixed-cone algorithms is not. Despite its practical advantages, the anti-k_t clustering procedure is somehow unrelated to the pattern of QCD radiation, which was one of the motivations for introducing sequential algorithms. For instance, one cannot introduce a jet resolution parameter based on the anti-k_t distance to define jet rates, because this would be infrared unsafe. The standard procedure to define jet rates is to find all anti-k_t jets, and classify an event as having n jets, if only n jets have transverse momenta above a given threshold. For experimental analyses in which it is important to understand the substructure of jets (e.g. in boosted object searches, see chapter 4), other algorithms, like the Cambridge/Aachen, have been exploited. In fact, LHC experiments consider a variety of jet algorithms, and do not rely on anti-k_t only.

Another reason why cone algorithms, even if IRC unsafe, were preferred to sequential algorithms was the scaling of the latter with the number of particles. In fact, a naive implementation of sequential algorithms scales as N^3 with the number of initial particles N. The reason for this is that computing the minimum of the mutual distances between pairs of particles requires in general N^2 numbers, and this minimisation has to be performed N times, until no pseudo-particles are left. This scaling of sequential algorithms becomes impractical in busy environments such as hadron collisions.

While it is not possible to reduce the number of minimisations, one does not need to scan over all pairs of particles to find the minimum of the generalised-k_t distances. In fact, the particles p_i and p_j that have the smallest distance need to be nearest neighbours in the y–ϕ plane [21]. To show this, let us assume that $p_{ti}^p < p_{tj}^p$. If there exist another particle p_l such that $\Delta R_{il} < \Delta R_{ij}$, then necessarily $d_{il} = \min(p_{ti}^{2p}, p_{tl}^{2p})\Delta R_{il}^2/R^2 < d_{ij}$, in contradiction with the fact that d_{ij} is the minimum of the generalised-k_t distances. This finding triggered a huge improvement in the speed of implementations of sequential algorithms [21]. One can in fact use methods from computational geometry to look for nearest neighbours in the y–ϕ plane and reduce the scaling of sequential algorithms with the number of particles. For instance, the scaling of the k_t-algorithm can be reduced to $N \ln N$ [21], whereas, for anti-k_t algorithm, the same strategy gives at most a $N^{3/2}$ scaling [14]. Note that compiling a list of nearest neighbours and updating it already reduces the complexity of the algorithm to $\mathcal{O}(N^2)$. This scaling, independent of the value of p, is better than the ones obtained by using computational geometry for $N \lesssim 10000$, which is in general sufficient unless one considers heavy-ion collisions.

2.2 Common issues with jet algorithms

Having introduced the main families of jet algorithms, we will now devote an entire section to two issues that are common to all jet algorithms. The first is the problem of determining the correct 'jet energy scale', i.e. assessing the 'true' value of a jet's momentum out of what is observed in the detectors. This affects in particular jets in hadronic collisions. The second is the effect of the recombination scheme, the procedure used to determine the momentum of a jet out of the momenta of its components.

2.2.1 Jet energy scale and jet areas

Since jets are the main objects that enter physics analyses in hadron collisions, it is crucial to be able to reconstruct their momenta as accurately as possible from the information obtained from the detectors. In fact, interesting jet distributions, such as those in jet transverse momentum or invariant mass, fall steeply with the increase of these variables, so that a migration of events between bins of such distributions due to an incorrect assignment of the energy of a jet has a huge effect on their overall shapes, undermining their correct interpretation. For instance, a peak in the invariant mass of two jets might reveal the presence of a new particle that decays hadronically, like a Z', a heavier partner of the Z boson, decaying into a quark–antiquark pair. The effect of a mis-measurement of the transverse momentum of jets might result in a broadening of the peak, which therefore becomes indistinguishable from QCD jet production. Alternatively, a migration of events from one bin to another might result in a 'fake' peak that does not correspond to any new particle.

The first serious issue is the correct determination of the energy of a jet out of the corresponding calorimetric deposits, which is commonly referred to as the problem of 'jet energy scale'. In practice, one tries to find the correction factor to be applied to the observed transverse momentum (or transverse energy) of a jet, to get its actual value, the one to be used in physics analyses. This procedure, commonly referred to as 'calibration', is very complicated, has to be repeated for each jet algorithm, and is strongly dependent on the specific experimental set-up. This is why we will not attempt to describe all experimental procedures needed to perform jet calibration, for which the interested reader is referred to experimental notes (e.g. [8, 22–25]). Here we will instead highlight the main sources of uncertainties, and discuss in some detail one detector-independent issue, the removal of a uniform background.

From a purely experimental point of view, one needs to take into account, for instance: the segmentation of calorimeters, whose cells have a finite size, which might be different in the central and in the forward/backward regions; the availability of tracking information, i.e. the fact that the direction of charged particles can be measured only in a central region, and elsewhere only calorimetric information is available; noise in the detectors, as well as imperfections, like cracks, transition regions, or even faults; unstable particles, whose decay products might fall in different parts of the detector. These are just examples of the many issues that experiments have to face when performing jet calibration. Some of these problems can be tackled offline, for instance by sending a beam of hadrons of known energy against the detectors, and studying their response. Such offline tests have to be validated when the experiment is running. This is because individual detectors are surrounded by additional material, which modifies the signals each particle give rise to. One common procedure to determine the jet energy scale online consists in checking that the transverse momentum of a jet is the same as that of a well-measured object (e.g. a photon or a Z boson decaying into an electron–positron pair) recoiling against it.

Even if the momentum of jets were know with infinite precision, all jet observables in hadron collisions are contaminated by a large background that has nothing to do with the high-energy collisions one is interested in. A first source of background is the so-called pile-up (PU), secondary low-energy collisions that occur at every crossing of the beams. The size of this effect at a high-luminosity machine like the LHC can be appreciated from figure 2.10, where one can see the peak number of interactions per beam crossing, as recorded by the CMS detector as a function of time during the first runs of the LHC. These numbers increase in the latest run of the LHC, with $\langle n_{\mathrm{PU}} \rangle = 40$ PU events recorded for an instantaneous luminosity $\mathcal{L} = 4 \times 10^{34}\ \mathrm{cm}^{-2}\ \mathrm{s}^{-1}$ and as many as 200 PU events at $\mathcal{L} = 7.5 \times 10^{34}\ \mathrm{cm}^{-2}\ \mathrm{s}^{-1}$ [27]. Another source of background is the so-called 'underlying event' (UE), i.e. beam-remnant interactions. These happen because, in every hadronic collision, the remainders of the collision of the incoming hadrons are coloured particles, and hence can interact via the strong force. Only rarely can these interactions give rise to a secondary hard collision, this occurrence being referred to as 'double-parton scattering'. More commonly, the remnants undergo a number of low-energy collisions, producing a large number of soft hadrons, many of which can be observed in regions of the detectors where hard jets are typically tagged. These are known as the 'diffuse' component of the UE, whereas double-parton

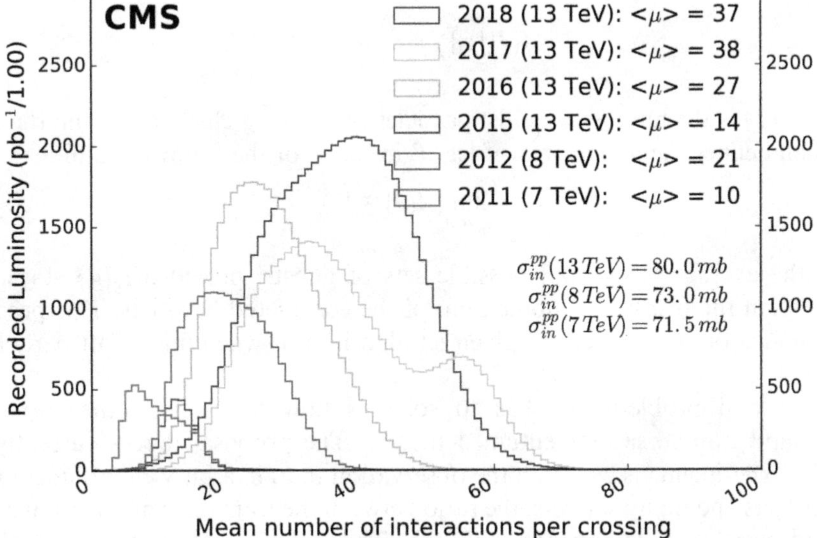

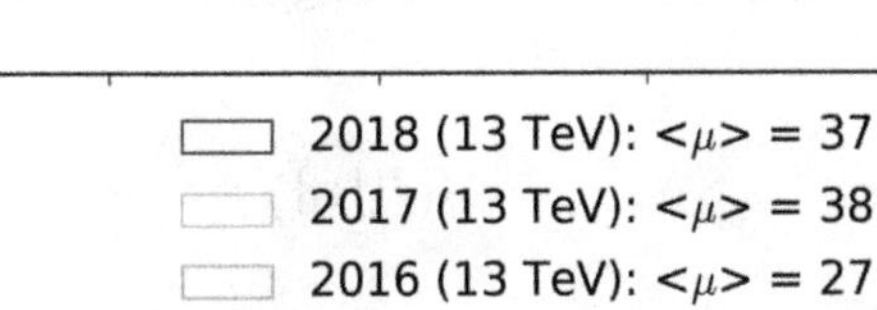

Figure 2.10. Number of interactions per beam crossing ($\langle\mu\rangle = \langle n_{\rm PU}\rangle$) recorded by the CMS experiment during the first two runs of the LHC [26]. Copyright &© 2008-2022 by the contributing authors.

scattering is referred to as its 'point-like' component. PU and UE have in common that they produce a yield of particles that is roughly uniform in rapidity and azimuth.

The effects of detector noise, PU and UE cannot be cleanly separated, and have to be dealt with concurrently. A problem that can be solved in a detector-independent way is the removal of a 'uniform' background, i.e. any sources of a fixed amount of transverse momentum ρ per unit rapidity and unit azimuth. If ρ is known with infinite precision, and each jet $p_{\rm J}$ is just a patch of fixed area A in the y–ϕ plane, we have

$$p_{t,\,\rm J}^{\rm true} \simeq p_{t,\,\rm J}^{\rm meas} - \rho A, \tag{2.7}$$

where $p_{t,\,\rm J}^{\rm true}$ is the actual value of the transverse momentum of the jet, and $p_{t,\,\rm J}^{\rm meas}$ its measured value[3]. In practice, however, ρ and A are not known *a priori*. Therefore, one needs to find a way to assess the sensitivity of a jet to a uniform background, and a sensible strategy to measure ρ.

A well-established means of quantifying a jet's sensitivity to a uniform background is the notion of 'active area' [28]. This is defined by generating a set g of ultra-soft 'ghost' particles $\{g_i\}$, with $\nu_{\rm g}$ ghost particles per unit area in the y–ϕ plane,

[3] For simplicity, we have neglected the fact that the transverse momentum is a two-dimensional vector. Therefore, the procedure in equation (2.7) needs to be improved to take into account how exactly the transverse momentum of a jet is constructed out of the momenta of its constituents.

and average transverse momentum $\langle g_t \rangle$, and considering, for each hard jet p_J, the quantity

$$A(J|\{g_i\}) = \frac{\mathcal{N}_g(J)}{\nu_g}, \tag{2.8}$$

where $\mathcal{N}_g(J)$ is the number of ghost particles of the set g clustered within the jet 'J'. One then defines the active area of jet 'J' in terms of the following limit:

$$A_J = \lim_{\nu_g \to \infty} \langle A(J|\{g_i\}) \rangle_g, \tag{2.9}$$

where the average is over all possible sets of ghosts, provided $\nu_g \langle g_t \rangle$ stays much smaller than the transverse momentum of the considered hard jets. This procedure gives an idea on how 'catchy' a given jet algorithm is when many soft particles are present.

The second problem one has to solve is how to estimate the size of the background transverse momentum density ρ. The proposal that is currently used by LHC experiments is based on the observation that, in a busy environment with a few hard jets and many soft jets, the ratio between the transverse momentum of most jets and their area is roughly constant. The only exception to this scaling is constituted by hard jets. In fact, if all jets were produced by a uniform background, the transverse momentum of each jet would be proportional to the jet area. Since in a regime with a high pile-up the number of soft jets is much larger than that of hard jets, one can use the following estimator [29]:

$$\rho = \text{median}_J \left(\frac{p_{tJ}}{A_J} \right), \tag{2.10}$$

and correspondingly find estimators for the standard deviation of ρ. Notice that, in the absence of PU, the above equation provides a measurement of the size of the diffuse component of the UE. Then one can obtain, on an event-by-event basis, the subtracted transverse momentum of a jet by replacing, in equation (2.7), the generic area A with A_J, the jet area defined in equation (2.9). The subtraction can be improved by taking into account the vectorial nature of transverse momentum [29]. Also, if performed before jet calibration, the procedure described above eliminates all detector noise that gives a uniform background.

We now look in more detail at the active area of the algorithms we have defined so far. We would expect that, for a jet of radius R, this area would be πR^2, or very close to that value. A closer look at each algorithm will reveal that this is generally not the case.

Let us consider first cone algorithms, and study the active area of isolated hard jets, e.g. separated in rapidity from other hard jets by more than twice the jet radius, defined with a seedless cone algorithm (like e.g. SISCone) with radius R, and a split-merge procedure to deal with overlapping cones. Simulations of dijet events in hadron collisions show that the active area of each hard jet is a mild function of the jet's transverse momentum, with an average around $\pi R^2/2$ (see the left-hand panel of

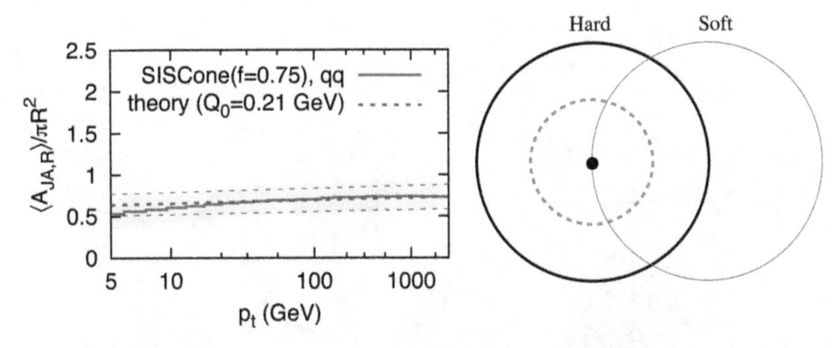

Figure 2.11. Left: the average area of a jet defined with the SISCone algorithm, as a function of the jet transverse momentum. Reproduced with permission from [28] Right: maximal overlap between a hard stable cone, ('Hard', black), and a stable cone made up of ghost particles ('Soft', blue). The picture, drawn by the author, is adapted from [28].

figure 2.11). The reason for this value relies on the split-merge procedure of cone algorithms, and is better understood by considering the example of an isolated stable cone containing a single hard parton [28]. If we add a uniform background of ghost particles, we obtain new soft stable cones made up of ghost particles only. The maximum overlap between the hard cone and a soft stable cone corresponds to the situation where the boundary of the soft cone touches the centre of the hard cone, as depicted in the right-hand panel of figure 2.11. From geometrical considerations, one finds that the fraction of particles that is contained in both cones is $f_{\text{max}} = \frac{2}{3} - \frac{\sqrt{3}}{2\pi} \simeq 0.391$, which is below the commonly chosen overlap thresholds $f = 0.5$ or $f = 0.75$. This means that the common particles will be assigned to either jet, according to which jet axis is closer. Indeed, if one considers all possible ghost stable cones that have a maximal overlap with the hard cone, only particles within a radius $R/2$ from the axis of the hard cone will be part of the hard jet, which will have now an active area $A_{\text{J}} = \pi R^2/4$. This value gets modified significantly in the presence of additional partons, and in the end the average area of hard jets is close to $\pi R^2/2$, with a mild increase with the jet transverse momentum.

Turning to sequential algorithms, the k_t algorithm clusters particles starting from the softest ones. As a result, the active area of jets reconstructed with this algorithm shows a greater dependence on the transverse momentum of the jets, as can be seen from the corresponding curves in figure 2.12, and large fluctuations (not shown in the figure). This is not the case of the anti-k_t algorithm, which confirms the intuitive picture of figure 2.9. The average active area of anti-k_t isolated hard jets is πR^2 with very small fluctuations, and basically no dependence on the jet transverse momentum [14]. The Cambridge/Aachen performs in a similar way to the anti-k_t, although with a larger dependence on the jet transverse momentum. Note that it is possible to have circular hard jets in the y–ϕ plane using cone algorithms as well, at the cost of abandoning the split-merge procedure. For instance, one can use SISCone to find all stable cones, and a progressive removal approach to deal with overlapping cones [30].

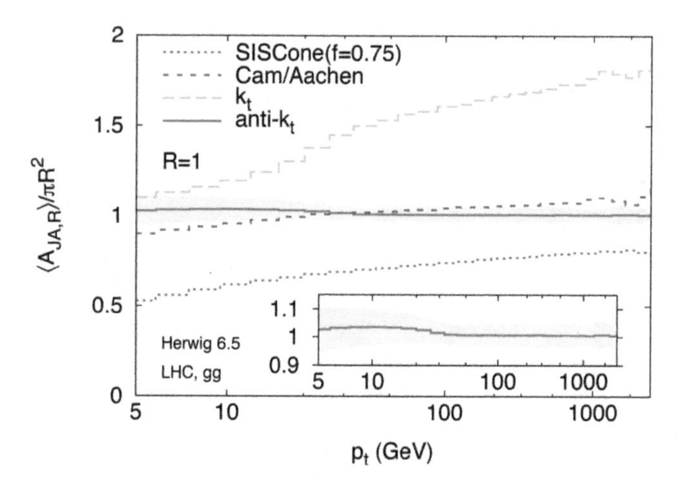

Figure 2.12. The mean value of the active area of the four algorithms described in the text, for jets initiated by gluons. Reproduced with permission from [14] © IOP Publishing. All rights reserved.

This procedure looks for the hardest stable cone and calls it a jet. That cone is then removed from the list of stable cones and its particles are removed from the list of particles. The procedure is then repeated until no stable cones are left. SISCone with progressive removal scales as $N^2 \ln N$ for N input particles [31].

Having the quantitative notion of the area of a jet, the procedure of subtracting a uniform background becomes possible on an event-by-event basis for any jet algorithm, not only for those whose jets have a fixed area. However, one needs to take into account that background is not completely uniform in rapidity. For instance, detector noise is different according to which parts of the experimental apparatus, in particular of the calorimeters, one considers. One can in principle devise rapidity-dependent estimators for ρ, for instance by dividing the y–ϕ plane in regions that are small enough to have a smooth function of y, but enough jets so that the estimate in equation (2.10) can be trusted. It is, however, more complicated to determine the correct jet energy scale for jets that cluster pseudo-particles in an unpredictable way, as happens for the k_t and Cambridge/Aachen algorithms. This is why the anti-k_t, whose jets are localised in the y–ϕ plane, is the preferred algorithm for measurements of jet observables.

The area method to remove a uniform background has been used in the LHC Run-I, when the average number of pile-up interactions was quite low, namely $\langle n_{\rm PU}\rangle = 20$. With higher pile-up, as at the LHC Run-II, and beyond, particle-level removal strategies are more effective. As the name suggests, these methods aim at determining whether each particle (or calorimetric deposit) arises from the primary interaction of a high-energy collision, or from secondary soft collisions. Then, jet algorithms are applied only to the particles that are assigned to the primary collision. These procedures, unlike the jet-area method, are not applied directly to jets. They also require some knowledge of the details of the detectors. Therefore,

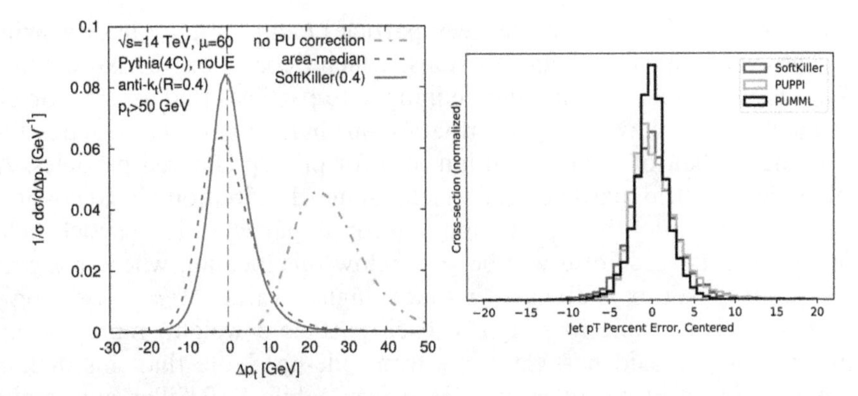

Figure 2.13. Differential cross section as a function of jet resolution for different PU mitigation procedures and different underlying processes. Left: dijet events corresponding to the parameters shown in the figure; the number 0.4 is the size of the patches in the y–ϕ plane needed by the SoftKiller procedure. © [32] (2015). With permission of Springer. Right: light-quark-initiated jets coming from the decay of a scalar with mass 500 GeV decaying into a $q\bar{q}$ pair at the LHC with $\sqrt{s}$ = 13 TeV. © [33] (2017). With permission of Springer.

we will only mention them very briefly, to give the reader an idea on how jets are constructed in practical applications. Also, some experimental analyses such as searches for boosted heavy particles which decay hadronically, require the use of jets with a large radius. In those cases, further 'grooming' procedures are used to clean such jets from softer particles. These will be discussed in more details in chapter 4.

A viable possibility to remove pile-up, which improves on the jet-area method, is to use information over charged tracks, as with current technology it is possible to assign each charged particle to the primary or a secondary interaction. Once PU/UE charged particles have been removed, the jet-area method can be applied to the remaining particles. A more drastic approach, named SoftKiller [32], is a fully fledged particle-level PU removal. Its inputs are square patches in the rapidity–azimuth plane. It removes all patches that have a transverse momentum smaller than $p_t^{\text{cut}} = \text{median}_{i\in\text{patches}}\{p_{ti}^{\max}\}$, where $p_{ti}^{\max}$ is the largest transverse momentum in the patch. From the definition, p_t^{cut} is the transverse momentum such that the procedure removes exactly half of the patches. For high pile-up, SoftKiller improves the resolution in jet transverse momentum, defined as the difference between the transverse momentum of a jet produced in a hard collision and its reconstructed transverse momentum after inclusion of pile-up (see the left-hand panel of figure 2.13). Another popular particle-level PU removal, developed at the same time as SoftKiller, is the Pile-Up per Particle Identification (PUPPI). This method is based on a shape variable, which for each particle i is defined as

$$\alpha_i = \ln \sum_j \frac{p_{tj}}{\Delta R_{ij}} \Theta(\Delta R_{ij} - R_{\min})\Theta(R_0 - \Delta R_{ij}), \tag{2.11}$$

where ΔR_{ij} is the usual distance between particles i and j in the rapidity–azimuth plane. The two Heaviside theta-functions, which depend on the parameters R_0 and R_{min}, limit the reach of the sum defining α_i to particles that are not too close, but also not too far away. Using information on charged particles, it is possible to compute the median of the distribution of α for pile-up charged particles. Also, assuming that neutral particles follow the same distribution, it is possible to construct the median of the distribution of α for all particles. If a particle belongs to pile-up, then its value of α will be well below this median, whereas a particle from a primary vertex will have a much higher value of α. Based on this observation, it is possible to assign to each particle a weight, representing the probability that the said particle arises from pile-up. Note that this method is quite similar in spirit to SoftKiller. However, while SoftKiller cuts emission irrespective of their position with respect to the hard jets, PUPPI retains information on the position of each particle in the rapidity–azimuth plane.

Both SoftKiller and PUPPI use criteria that are understandable by humans. In fact, pile-up removal is a very interesting avenue for the exploitation of machine-learning techniques, due to the fact that energy deposits in the rapidity–azimuth plane could be considered as images. These can then be filtered via a convolutional neural network that leaves only the particles assigned to the primary interaction. This is the idea behind the procedure of Pile-up Mitigation with Machine Learning (PUMML) proposed in [33]. PUMML gives a further improvement with respect to SoftKiller and PUPPI at very high pile-up, as can be seen for instance in the distribution of the jet-p_t resolution for $\langle n_{\text{PU}} \rangle = 140$ in the right-hand panel of figure 2.13. The authors of [33] comment also on what discrimination procedure the neural network actually learns. In particular, they look at the functional relation between the actual transverse momentum of a jet produced in a primary interaction, with respect to the measured transverse momentum and the typical pile-up transverse momentum. They find that PUMML acts in a similar way to another human procedure for pile-up removal called jet cleansing [34]. The example of PUMML shows how the use of machine learning in jet physics is valuably enhanced by physics intuition developed in devising simple procedures, with artificial intelligence stepping in to refine and optimise such procedures. Along these lines, the authors of [35] proposes a refinement of PUPPI using gated graph neural networks [36], called PUPPI Machine Learning (PUPPIML). Without going into the details, this refinement considers more variables with respect to PUMML, and massively exploits the details of the detectors. It is instructive to look at figure 2.14 to have a qualitative picture of how different pile-up procedures work. There one can see different events in the rapidity–azimuth plane. The left-most panels correspond to the particles produced in a hard primary collision. Moving to the right, pile-up is added, and one can immediately see the amount of noise induced by pile-up. Looking at the other panels to the right, corresponding to PUPPI, SoftKiller and PUPPIML, one observes that all methods are able to remove most of the particles originating from pile-up. Tiny differences appear, for instance PUPPI seems to keep pile-up particles closer to the hardest particles in the event, whereas SoftKiller on the contrary keeps particles that are far away from the hard jets. Both are discarded by PUPPIML, which acts a refinement of both procedures.

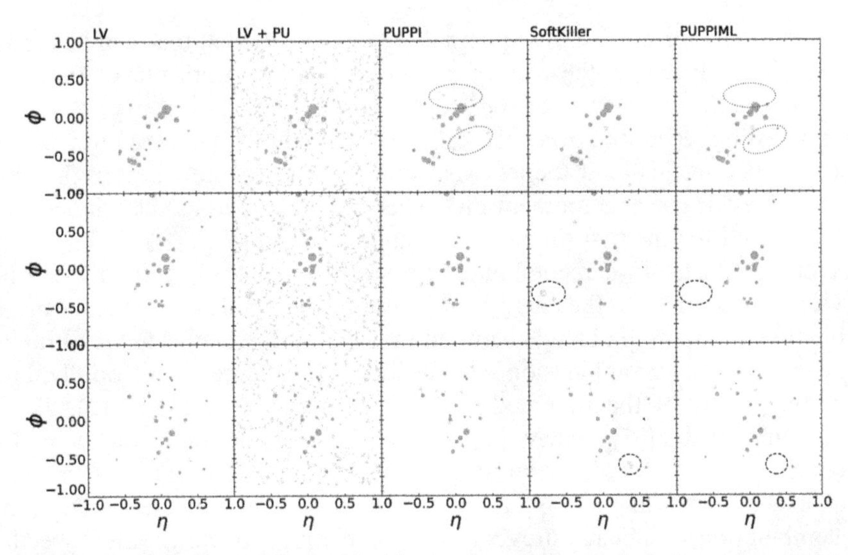

Figure 2.14. The performance of different PU mitigation procedures on a single three-jet event at the LHC with $\sqrt{s}$ = 13 TeV. Reproduced from [35]. Copyright © 2019, Società Italiana di Fisica and Springer-Verlag GmbH Germany, part of Springer Nature.

2.2.2 Recombination schemes

We conclude this section by discussing various procedures that can be used in a sequential algorithm to merge two particles into a single pseudo-particle. These procedures are known as 'recombination schemes'. Given two pseudo-particles, a recombination scheme determines how the momenta p_i and p_j of two pseudo-particles have to be recombined into a new pseudo-particle of momentum p_{ij}. Three recombination schemes use the particles' energy and three-momenta, as follows:

- *E*-scheme: $p_{ij} = p_i + p_j$, i.e. addition of pseudo-particles' four-momenta;
- *E*0-scheme: $E_{ij} = E_i + E_j$, and $\vec{p}_{ij} = (\vec{p}_i + \vec{p}_j)/E_{ij}$, so that the resulting jet is massless.
- *P*-scheme: $\vec{p}_{ij} = \vec{p}_i + \vec{p}_j$, and $E_{ij} = |\vec{p}_{ij}|$, so as to have again a massless jet.

Other recombination schemes determine the transverse momentum $p_{t,ij}$, the rapidity y_{ij} and the azimuthal angle ϕ_{ij} of the new pseudo-particle out of the transverse momenta, rapidities and azimuths of the parent pseudo-particles [3]. They can all be obtained from the relations

$$p_{t,ij} = p_{ti} + p_{tj}, \quad y_{ij} = \frac{p_{ti}^p y_i + p_{tj}^p y_j}{p_{ti}^p + p_{tj}^p}, \quad \phi_{ij} = \frac{p_{ti}^p \phi_i + p_{tj}^p \phi_j}{p_{ti}^p + p_{tj}^p}, \tag{2.12}$$

with p a positive parameter. More recombination schemes can be obtained by replacing transverse momenta with transverse energies, and rapidities with pseudorapidities. The above schemes can be generalised to any number of recombined particles, and are not specific to sequential algorithms. In fact, they can be used with cone algorithm to obtain the axis of a cone out of the momenta of the pseudo-particles inside it.

An interesting recombination procedure is obtained by taking the limit $p \to \infty$ in equation (2.12). In this limit the jet axis coincides with the direction of the particle with the largest transverse momentum[4]. This axis is known as the winner-take-all axis, introduced for the first time in e^+e^- annihilation in [37].

One of the effects of the recombination scheme is that of changing the sensitivity of physical observables to the energy–momentum flow inside each jet. For instance, variables like the azimuthal angle between two jets are insensitive to QCD radiation inside each jet if the recombination scheme adds three-momenta vectorially, but not if it performs any of the weighted recombinations in equation (2.12). This was pointed out first in [38], where a p_t-weighted recombination with $p = 1$ was considered, and later for the azimuthal correlation of a Z boson and the winner-take-all axis of a recoiling jet [39]. Although very interesting theoretically, changing the recombination scheme requires additional experimental work. This is specifically due to the fact that the transverse momentum of a jet depends on the recombination procedure, so that jet energy-scale calibration has to be repeated for each recombination procedure within the same jet algorithm. Since jet calibration is a complicated procedure, experiments select not only a default jet algorithm, but also a default recombination scheme. For instance, the LHC default is the anti-k_t algorithm with the E recombination scheme. Note that the two main LHC experiments, ATLAS and CMS, adopt slightly different choices for the jet radius, so as to exploit the sensitivity of their detectors as much as possible [22, 40]. It has to be stressed that, practical issues aside, there is nothing that prevents an experiment from using a different jet algorithm for a specific analyses, or exploring a different recombination scheme within the same jet algorithm.

The discussion so far has shown how difficult it is to move from the intuitive concept of a jet as a huge deposit of energy in a detector to its rigorous definition in terms of an algorithmic procedure. The algorithms we have described so far are the ones that have been actually used in high-energy physics experiments. Their properties have been thoroughly tested, and are well understood. In recent years however, especially to make better use of jets as tools to discover new particles, novel and sometimes unconventional ideas to reconstruct jets have been proposed. The next section is devoted to presenting a number of these recent developments, so as to give the reader a sense of how the field might evolve in the future.

2.3 Further jet clustering procedures

The algorithms that we have presented so far are those that have been most widely used by experiments. Nevertheless, jet algorithms is an evolving field, with more and more algorithms being devised. Many of them are publicly available as plugins to

[4] Note that, in order to ensure the collinear safety of the recombination procedure, the transverse momentum of the jet obtained from equation (2.12) has to depend *linearly* on the transverse momentum of the recombined particles.

the computer library FASTJET [31], whose manual contains concise description of the implemented algorithms. Instead of giving an overview of all most recent developments, here we will present ideas on how jets could be reconstructed that do not fall immediately into the category of cone and sequential algorithms. The examples we have chosen aim at highlighting various complementary ways to approach the problem of finding and characterising jets.

Inclusive jet algorithms in e^+e^- annihilation. The working of sequential jet algorithm for hadron collisions that we have presented in section 2.1.2 is referred to as 'inclusive mode'. There is, however, the possibility of running generalised k_t-algorithms with positive power p in 'exclusive mode', similar to what is done for the Durham algorithm in e^+e^- annihilation. In particular, one fixes a jet resolution d_{cut}, and each event is classified as having n-jets if and only if the minimum of the mutual distances d_{ij} and of the distance of each particle with the beam d_{iB} is larger than d_{cut}. This creates an event with n pseudo-particles, and all the particles that are not clustered with the jets are part of a 'beam' jet. Each particle clustered with the beam jet would be considered as a jet, were the algorithm run in inclusive mode.

All algorithms we have presented for e^+e^- do not cluster particles with the beam, as this gives just an arbitrary direction for strongly interacting particles. Nevertheless, it might be very useful to have jet algorithms defined in e^+e^- annihilation that mimic those for hadron collisions. This opens for instance the possibility of testing the performance of various algorithms in an experimentally clean environment. First we need to understand the relation between distance measures in e^+e^- and hadron collisions. To do this, let us consider, at a hadron collider, two particles that are quasi-central, i.e. whose rapidity is very close to zero. The energy of these particles is almost equal to their transverse momentum. In particular, considering two particles p_i and p_j that are very close in angle, we can approximate

$$2(p_i p_j) = 2E_i E_j(1 - \cos\theta_{ij}) \simeq p_{ti} p_{tj} \Delta R_{ij}^2. \tag{2.13}$$

Therefore, for collinear particles, we can identify ΔR_{ij}^2 with the angular distance $2(1 - \cos\theta_{ij}) \simeq \theta_{ij}^2$, up to a finite, rapidity-dependent overall factor. Therefore, a suitable definition of generalised-k_t distance in e^+e^- annihilation is given by

$$d_{ij} = \min(E_i^{2p}, E_j^{2p}) \frac{1 - \cos\theta_{ij}}{1 - \cos R}. \tag{2.14}$$

We can then exploit the correspondence between transverse momentum and energy to introduce a distance with a 'beam', which is simply given by $d_{i\text{B}} = E_i^{2p}$ [31, 41]. With such distances, the algorithm works as in hadron collisions: if, at any step, the minimum of all distances is $d_{i\text{B}}$, pseudo-particle p_i is considered as a jet and removed from the list of pseudo-particles. If the minimum distance is d_{ij}, pseudo-particles p_i and p_j are recombined. The procedure stops when no pseudo-particles are left. Similarly, it is also possible to devise an e^+e^- version of SISCone [41]. With these adaptations, one can study the behaviour of jet algorithms in a simpler environment than hadron collisions. For instance, in e^+e^- annihilation it is possible to compute jet rates in QCD with high precision [41], or even gain an analytical understanding of the properties of various algorithms, and exploit this knowledge to devise better algorithms at hadron colliders [42].

Quantum jets. As explained at the beginning of this chapter, the main purpose of a jet algorithm is that of mapping a set of final-state particles into a small number of jets, whose momenta should be close to those of the primordial hard quarks and gluons produced in a high-energy collision. This perfect match is only guaranteed in a case where the initial hard partons are accompanied by a set of infinitely soft emissions and collinear splittings. In real life, it might happen that the number of jets an event is mapped into does not reflect the original hard event, and moreover that this result depends on the jet algorithm. A notion that emphasises this intrinsic uncertainty in the interpretation of events is that of 'quantum jets', or simply 'Q-jets' [43]. They are constructed out of a 'quantum' jet algorithm, in which an event is not mapped into a number of jets with certainty, but with a given probability. More specifically, a quantum jet algorithms is similar to a 'classical' sequential algorithm, with the following modifications occurring at each stage:

1. A weight ω_{ij} is computed for every pair of pseudo-particles p_i and p_j.
2. A pair p_i and p_j is chosen with probability $\Omega_{ij} = \omega_{ij}/\sum_{k<l}\omega_{kl}$, and merged into a single pseudo-particle.

The classical procedure is recovered if $\omega_{ij} = \delta(d_{ij} - d_{\min})$, where d_{ij} is an arbitrary IRC safe distance measure, and $d_{\min}$ is its smallest value among all possible pairs of pseudo-particles. The procedure is repeated N_{tree} times, each time giving a 'tree' of clusterings, the probability of each tree being $\prod_{\text{mergings}} \Omega_{ij}$. A particularly interesting class of weights is

$$\omega_{ij} = \exp\left[-\alpha\frac{d_{ij} - d_{\min}}{d_{\min}}\right] \tag{2.15}$$

with α a positive number called 'rigidity'. Notice that, for $\alpha \to \infty$ the pair with the minimum d_{ij} will always be chosen, thus recovering the corresponding classical sequential algorithm. This is very close to what happens in quantum mechanics, where quantum trajectories are distributed according to a probability that is peaked around classical trajectories, with the role of the parameter α played by the constant $1/\hbar$. In practice, each event will not be clustered unambiguously into jets, but we will obtain rather N_{tree} different sets of jet momenta, each with a different assigned probability. The trees constructed out of this probabilistic interpretation are called Q-jets. Despite the idea being quite innovative in spirit, Q-jets have not been massively used in experimental analyses. Their applications are restricted to boosted object searches, and will therefore be discussed in chapter 4.

Jet-finding as an optimisation procedure. Let us consider, as a starting point, a two-jet event in e^+e^- annihilation. In line with the intuitive idea of a jet as a region where energy deposit is maximised, one can find the direction that maximises the scalar sum of the projection of momenta along it. This maximum is called the thrust

$$T = \max_{\vec{n}_T}\frac{\sum_i|\vec{p}_i \cdot \vec{n}_T|}{\sum_i|\vec{p}_i|}, \tag{2.16}$$

and the direction $\vec{n}_{\mathrm{T}}$ maximising the sum in the above equation is called the thrust axis. In e^+e^- annihilation, for each event, one can cluster all particles into two back-to-back jets as follows: the thrust axis can be identified with the jet axis, and particles can be assigned to either jet according to whether $\vec{p}_i \cdot \vec{n}_{\mathrm{T}}$ is larger or smaller than zero. An analogous procedure, generalisable to an arbitrary jet configuration, is to construct a quantity called 2-jettiness $\mathcal{T}_2(n_1, n_2)$. Given two light-like vectors $n_i = (1, \vec{n}_i)$ (which of course implies $\vec{n}_i$ is a unit three-vector, i.e. it specifies a direction), the 2-jettiness is defined as

$$\mathcal{T}_2(n_1, n_2) \equiv \frac{1}{\sum_i |\vec{p}_i|} \sum_i \min\{2n_1 \cdot p_i, 2n_2 \cdot p_i\}. \tag{2.17}$$

To define a 'beam jet' collecting all particles that are not clustered with the jets we introduce a jet radius R and redefine the 2-jettiness as follows:

$$\mathcal{T}_2(n_1, n_2) \equiv \frac{1}{\sum_i |\vec{p}_i|} \sum_i \min\left\{E_i, \frac{2n_1 \cdot p_i}{R^2}, \frac{2n_2 \cdot p_i}{R^2}\right\}. \tag{2.18}$$

We can now use the 2-jettiness as a jet algorithm, by minimising its value over all possible light-like vectors n_1, n_2. The unit vectors $\vec{n}_1$ and $\vec{n}_2$ that correspond to the minimum of $\mathcal{T}_2(n_1, n_2)$ define the jet directions. Once the two directions $\vec{n}_1$ and $\vec{n}_2$ have been found, particle p_i is assigned to either jet, or to the beam jet, according to which is the minimum between $2(n_1 p_i)/R^2$, $2(n_2 p_i)/R^2$ and E_i. If n_1 and n_2 are two back-to-back directions, i.e. $n_2 = (1, -\vec{n}_1)$, and $R = 1$, the minimum of the 2-jettiness is essentially one minus the thrust in equation (2.16).

Similarly, one can find N jets $\{n_1, n_2, \ldots, n_N\}$ by minimising N-jettiness, defined as [44, 45]

$$\mathcal{T}_N(n_1, n_2, \ldots, n_N) = \frac{1}{\sum_i |\vec{p}_i|} \sum_i \min\left\{E_i, \frac{2n_1 \cdot p_i}{R^2}, \frac{2n_2 \cdot p_i}{R^2}, \ldots, \frac{2n_N \cdot p_i}{R^2}\right\}. \tag{2.19}$$

The above definition of N-jettiness can be also modified by allowing different distance measures (or 'metrics'), both in e^+e^- and in hadron collisions. In particular, in hadron collisions, choosing the distance between a particle p_i and a direction n_A as $2\cosh y_a(n_A p_i)/R^2$ and that with the beam as the particle's transverse momentum p_{ti} leads to jet boundaries that are circles in the rapidity–azimuth plane. This gives rise to the *exclusive* cone algorithm XCone [45]. This algorithm is exclusive in that the number of jets is fixed from the start, and $\mathcal{T}_N$ evaluated on the jet directions is the corresponding resolution variable.

A further generalisation of this procedure leads to an inspiring geometric interpretation of jet algorithms and related observables [46]. Given a set of particles with energies E_i and directions identified by the unit vectors $\vec{n}_i$ (with corresponding four-momenta $p_i = E_i(1, \vec{n}_i)$), we define the energy flow

$$\mathcal{E}(\vec{n}) \equiv \sum_i E_i \delta(\vec{n} - \vec{n}_i). \tag{2.20}$$

We then define the energy mover's distance (EMD) between two energy flows $\mathcal{E}$ (with M particles of momenta p_i) and $\mathcal{E}'$ (with M' particles of momenta p_j') as [47]

$$\mathrm{EMD}_{\beta,R}(\mathcal{E}, \mathcal{E}') \equiv \min_{f_{ij}\geqslant 0} \sum_{i=1}^{M} \sum_{j=1}^{M'} f_{ij} \left(\frac{\theta_{ij}}{R}\right)^{\beta} + \left|\sum_{i=1}^{M} E_i - \sum_{j=1}^{M'} E_j'\right|, \tag{2.21}$$

with $\theta_{ij} \equiv \sqrt{1 - \vec{n}_i \cdot \vec{n}_j}$ with f_{ij} satisfying the constraints

$$\sum_{i=1}^{M} f_{ij} \leqslant E'_j, \quad \sum_{j=1}^{M'} f_{ij} \leqslant E_i, \quad \sum_{i=1}^{M}\sum_{j=1}^{M'} f_{ij} = \min\left(\sum_{i=1}^{M} E_i, \sum_{j=1}^{M'} E'_j\right). \tag{2.22}$$

It turns out that many jet algorithms and related observables can be expressed in terms suitable EMDs. Let us start by considering one minus the thrust. This is obtained by minimising an EMD between a generic energy flow $\mathcal{E}$ and another energy flow $\mathcal{E}'$ that belongs to a particular manifold. This manifold, which we call $\mathcal{P}_2^{\mathrm{BB}}$ contains all energy flows with two back-to-back particles with the same total energy as the event $\mathcal{E}$. Let a back-to-back direction be identified by the vector $\vec{n}$. If one starts from EMD is that of equation (2.21) with $\beta = 2$ and $f_{ij} \sim E_i E'_j$, and minimises it over all possible event flows in $\mathcal{P}_2^{\mathrm{BB}}$, one obtains

$$\min_{\mathcal{E}' \in \mathcal{P}_2^{\mathrm{BB}}} \mathrm{EMD}_2(\mathcal{E}, \mathcal{E}') = 2 \min_{\vec{n}} \sum_i E_i \min\{1 - \vec{n}_i \cdot \vec{n}, 1 + \vec{n}_i \cdot \vec{n}\}. \tag{2.23}$$

But the above equation, modulo a normalisation, is just one minus the thrust. Similarly, N-jettiness is the EMD between an energy flow $\mathcal{E}$ and an energy flow $\mathcal{E}'$ belonging to $\mathcal{P}_N$, the set of all N-particle energy flows. Finding the directions that minimise N-jettiness corresponds to finding N jets with the XCone algorithm, with the parameters of the EMD determining the jet boundaries. Note that, imposing that $\mathcal{E}'$ has the same total energy as $\mathcal{E}$ results in all particles being clustered within the jets. A beam jet can be enabled by just relaxing that constraint, and allowing $\mathcal{E}$ and $\mathcal{E}'$ to have different total energies.

It is also possible to recast each step of a sequential algorithm in this geometric framework. In fact, given an energy flow $\mathcal{E} \in \mathcal{P}_M$, the next step of a sequential algorithm will be the closest energy flow $\mathcal{E} \in \mathcal{P}_{M-1}$, according to a selected EMD. Remarkably, it can be shown mathematically that the closest energy flow is obtained only by merging two particles into one, or eliminating one of the particles, which is exactly how a sequential algorithm operates. Note that, for a given value of the parameters β and R in equation (2.21), the EMD minimisation procedure gives both a distance measure and a recombination scheme. For instance, $\beta = 1$ corresponds to the k_t-algorithm with the WTA scheme. Similarly, raising the distance to the power $1/\beta$ and taking the limit $\beta \to \infty$ corresponds to the Cambridge–Aachen algorithm, while there no EMD procedure that gives the anti-k_t has been found so far.

Remarkably, the subtraction of a uniform background can also be interpreted in terms of the minimisation of an EMD. In fact, let us consider an isotropic energy flow $\mathcal{U}$. The event $\mathcal{E}' + \rho\,\mathcal{U}$ corresponds to the contamination of the energy flow $\mathcal{U}$ with uniform background. Given an energy flow $\mathcal{E}$, we can minimise EMD $_\beta(\mathcal{E}, \mathcal{E}' + \rho\,\mathcal{U})$ over all possible event flows $\mathcal{E}'$. The energy flow $\mathcal{E}_C(\mathcal{E}, \rho)$ corresponding to that minimum is the energy flow $\mathcal{E}$ where a uniform background has been subtracted.

Recasting jet observables and algorithms in terms of an optimisation procedure opens up the possibility of analysing them from a mathematical point of view as optimal transport problems (see e.g. [48]), as well to recast them as machine learning problems [49], where artificial intelligence is exploited to achieve improved performance on the EMD optimisation.

There are many lessons that one can learn from the above discussion. An important one is that there can be different procedures that lead to qualitatively similar jets. It is then only a matter of finding which one makes it possible to achieve the goals it was designed for in the fastest and most efficient way. Exploring different jet-finding philosophies makes it easier to find new procedures that can lead to jets that are qualitatively different from the already known ones. This in turn can lead to even more possibilities to exploit jets as tools for high-energy physics.

References

[1] https://aleph.web.cern.ch/aleph/aleph/newpub/dali-displays.html

[2] Sterman G F and Weinberg S 1977 Jets from quantum chromodynamics *Phys. Rev. Lett.* **39** 1436

[3] Huth J E *et al* 1990 Toward a standardization of jet definitions *1990 DPF Summer Study on High-energy Physics: Research Directions for the Decade (Snowmass 90)* pp 134–6

[4] https://twiki.cern.ch/twiki/bin/view/AtlasPublic/EventDisplayPublicResults

[5] Arnison G *et al* (UA1 Collaboration) 1983 Hadronic jet production at the CERN proton–anti-proton collider *Phys. Lett.* B **132** 214

[6] Abe F *et al* (CDF Collaboration) 1992 The topology of three jet events in $\bar{p}p$ collisions at $\sqrt{s} = 1.8$ TeV *Phys. Rev.* D **45** 1448–58

[7] Seymour M H 1998 Jet shapes in hadron collisions: higher orders, resummation and hadronization *Nucl. Phys.* B **513** 269–300

[8] Blazey G C *et al* 2000 2000 Run II Jet Physics: Proc. Run II: QCD and Weak Boson Physics Workshop: Final General Meeting pp 547–77

[9] Salam G P and Soyez G 2007 A practical seedless infrared-safe cone jet algorithm *J. High Energy Phys.* **05** 086

[10] Kidonakis N, Oderda G and Sterman G F 1998 Evolution of color exchange in QCD hard scattering *Nucl. Phys.* B **531** 365–402

[11] Seymour M H and Tevlin C 2006 A comparison of two different jet algorithms for the top mass reconstruction at the LHC *J. High Energy Phys.* **11** 052

[12] Ellis S D, Huston J and Tonnesmann M 2001 On building better cone jet algorithms *eConf* **C010630** 513

[13] Albrow M G *et al* (TeV4LHC QCD Working Group Collaboration) 2006 *Tevatron-for-LHC Report of the QCD Working Group* p 10

[14] Cacciari M, Salam G P and Soyez G 2008 The anti-k_t jet clustering algorithm *J. High Energy Phys.* **04** 063

[15] Bartel W *et al* (JADE Collaboration) 1986 Experimental studies on multi-jet production in $e^+ e^-$ annihilation at PETRA energies *Z. Phys.* C **33** 23

[16] Catani S, Dokshitzer Y L, Olsson M, Turnock G and Webber B R 1991 New clustering algorithm for multi-jet cross-sections in $e^+ e^-$ annihilation *Phys. Lett.* B **269** 432–8

[17] Dokshitzer Y L, Leder G D, Moretti S and Webber B R 1997 Better jet clustering algorithms *J. High Energy Phys.* **08** 001

[18] Catani S, Dokshitzer Y L, Seymour M H and Webber B R 1993 Longitudinally invariant K_t clustering algorithms for hadron hadron collisions *Nucl. Phys.* B **406** 187–224

[19] Wobisch M and Wengler T 1998 Hadronization corrections to jet cross-sections in deep inelastic scattering *Workshop on Monte Carlo Generators for HERA Physics (Plenary Starting Meeting)* vol 4 pp 270–9
[20] Ellis S D and Soper D E 1993 Successive combination jet algorithm for hadron collisions *Phys. Rev.* D **48** 3160–6
[21] Cacciari M and Salam G P 2006 Dispelling the N^3 myth for the k_t jet-finder *Phys. Lett.* B **641** 57–61
[22] CMS Collaboration, Jet performance in pp collisions at 7 TeV https://cds.cern.ch/record/1279362
[23] Aad G *et al* (ATLAS Collaboration) 2013 Jet energy measurement with the ATLAS detector in proton–proton collisions at $\sqrt{s} = 7$ TeV *Eur. Phys. J.* C **73** 2304
[24] CMS Collaboration, Jet energy scale and resolution measurement with Run 2 Legacy Data Collected by CMS at 13 TeV https://cds.cern.ch/record/2792322
[25] Aad G *et al* (ATLAS Collaboration) 2021 Jet energy scale and resolution measured in proton–proton collisions at $\sqrt{s} = 13$ TeV with the ATLAS detector *Eur. Phys. J.* C **81** 689
[26] https://twiki.cern.ch/twiki/bin/view/CMSPublic/LumiPublicResults#Multi_year_plots
[27] Buffat X *et al* 2022 HL-LHC Experiment Data Quality Working Group Summary Report *Technical Report* CERN, Geneva, https://cds.cern.ch/record/2802720
[28] Cacciari M, Salam G P and Soyez G 2008 The catchment area of jets *J. High Energy Phys.* **04** 005
[29] Cacciari M and Salam G P 2008 Pileup subtraction using jet areas *Phys. Lett.* B **659** 119–26
[30] http://fastjet.fr/
[31] Cacciari M, Salam G P and Soyez G 2012 FastJet user manual *Eur. Phys. J.* C **72** 1896
[32] Cacciari M, Salam G P and Soyez G 2015 SoftKiller, a particle-level pileup removal method *Eur. Phys. J.* C **75** 59
[33] Komiske P T, Metodiev E M, Nachman B and Schwartz M D 2017 Pileup mitigation with machine learning (PUMML) *J. High Energy Phys.* **12** 051
[34] Krohn D, Schwartz M D, Low M and Wang L-T 2014 Jet cleansing: pileup removal at high luminosity *Phys. Rev.* D **90** 065020
[35] Arjona Martínez J, Cerri O, Pierini M, Spiropulu M and Vlimant J-R 2019 Pileup mitigation at the large hadron collider with graph neural networks *Eur. Phys. J. Plus* **134** 333
[36] Li Y, Tarlow D, Brockschmidt M and Zemel R 2015 Gated graph sequence neural networks *ICLR 2016*
[37] Larkoski A J, Neill D and Thaler J 2014 Jet shapes with the broadening axis *J. High Energy Phys.* **4** 17
[38] Banfi A, Dasgupta M and Delenda Y 2008 Azimuthal decorrelations between QCD jets at all orders *Phys. Lett.* B **665** 86–91
[39] Chien Y-T, Rahn R, Schrijnder van Velzen S, Shao D Y, Waalewijn W J and Wu B 2021 Recoil-free azimuthal angle for precision boson-jet correlation *Phys. Lett.* B **815** 136124
[40] Aad G *et al* (ATLAS Collaboration) 2009 Expected performance of the ATLAS experiment—detector, trigger and physics
[41] Weinzierl S 1565 Jet algorithms in electron–positron annihilation: perturbative higher order predictions *Eur. Phys. J.* C **71** 2011 [Erratum: 2011 *Eur. Phys. J.* C **71** 1717]
[42] Dasgupta M, Fregoso A, Marzani S and Powling A 2013 Jet substructure with analytical methods *Eur. Phys. J.* C **73** 2623

[43] Ellis S D, Hornig A, Roy T S, Krohn D and Schwartz M D 2012 Qjets: a non-deterministic approach to tree-based jet substructure *Phys. Rev. Lett.* **108** 182003

[44] Thaler J 2015 Jet maximization, axis minimization, and stable cone finding *Phys. Rev.* D **92** 074001

[45] Stewart I W, Tackmann F J, Thaler J, Vermilion C K and Wilkason T F 2015 XCone: N-jettiness as an exclusive cone jet algorithm *J. High Energy Phys.* **11** 072

[46] Komiske P T, Metodiev E M and Thaler J 2020 The hidden geometry of particle collisions *J. High Energy Phys.* **07** 006

[47] Komiske P T, Metodiev E M and Thaler J 2019 Metric space of collider events *Phys. Rev. Lett.* **123** 041801

[48] Peleg S, Werman M and Rom H 1989 A unified approach to the change of resolution: space and gray-level *IEEE Trans. Pattern Anal. Mach. Intell.* **11** 739–42

[49] Radovic A, Williams M, Rousseau D, Kagan M, Bonacorsi D, Himmel A, Aurisano A, Terao K and Wongjirad T 2018 Machine learning at the energy and intensity frontiers of particle physics *Nature* **560** 41–8

IOP Publishing

Hadronic Jets (Second Edition)
An introduction
Andrea Banfi

Chapter 3

QCD for jet physics

One the most amazing features of jet physics is that, despite jets being complicated objects constructed out of many hadrons, their basic properties can be understood in terms of the elementary degrees of freedom of quantum chromodynamics (QCD), i.e. quarks and gluons. A formal introduction to QCD is beyond the scope of this book, and is now textbook material [1, 2]. Here we will review the aspects of QCD that are crucial to understand the dynamics of jets.

QCD is a quantum field theory describing the interactions of elementary fermions, the quarks, mediated by spin-1 gauge particles, the gluons. As explained in the introduction, the known quarks are six, organised in three families. Their properties are summarised in table 1.1.

Besides electric charge (and weak isospin), each quark possesses an additional conserved charge, the colour. More specifically, each quark has three colours (say red, green and blue). Coloured particles, such as quarks, can interact via the strong force in a similar way to which electrically charged particles interact via the electromagnetic force. The strong force is mediated by spin-1 particles, the gluons, like the electromagnetic force is mediated by photons. There is, however, a fundamental difference between the electromagnetic and the strong force. When an electron emits a photon, it does not change its electric charge, whereas a quark emitting a gluon changes its colour. This means that gluons can take away colour and are colour-charged themselves, whereas photons are electrically neutral. Specifically, gluons carry eight different colours. Therefore, while in quantum electrodynamics (QED) the transition amplitude for an electron emitting a photon is only proportional to the electron charge ($-e$, where e is the electric charge of the proton), in QCD one needs to consider the transition amplitude for a quark of a colour j (an index between 1 and 3) transforming into a quark of a colour i (another index between 1 and 3) after the emission of a gluon of colour a (an index between 1 and 8). Any such QCD amplitude depends then not only on the strong interaction coupling constant g_s (the analogue of e), but also on a colour matrix t_{ij}^a. For quarks

doi:10.1088/978-0-7503-4737-2ch3

emitting gluons we have eight colour matrices, which form a Lie algebra under commutation, namely:

$$[t^a, t^b] = if^{abc}t^c. \tag{3.1}$$

These commutation relations are those of the generators of the Lie group $SU(3)$, which is in fact the gauge group of QCD. The symbol f^{abc}, totally antisymmetric in the indices a, b, c, embodies the so-called 'structure constants' of the group $SU(3)$.

Before discussing jet formation, we need to address the fact that no coloured particles are observed in our detectors. Despite this, for high-energy processes, we can compute hadronic cross sections using transition probabilities between unobserved quarks and gluons, and surprisingly we obtain extremely good agreement with experimental data. The key property that makes QCD predictive is asymptotic freedom. The strength of the interactions in QCD is determined by the strong coupling $\alpha_s = g_s^2/(4\pi)$, the analogue of the fine structure constant $\alpha = e^2/(4\pi)$ in QED. As in any quantum field theory, the value of α_s that rules the magnitude of a given transition probability depends on the typical scale of that transition. For instance, the total production rate of hadrons in electron–positron annihilation depends on $\alpha_s(\sqrt{s})$, with $\sqrt{s}$ the centre-of-mass energy of the electron–positron collision. The coupling α_s decreases for increasing momenta, and vanishes for asymptotically large momenta, which implies that quarks and gluons at high energy are essentially non-interacting particles. QED has the opposite behaviour, with the fine structure constant tending to the fixed value 1/137 for vanishingly small momenta. What happens at low momenta for QCD? The coupling α_s grows, until a point at which we cannot compute any observable in terms of quarks and gluons anymore. In fact, unless rare exceptions, in quantum field theory we can calculate only small deviations from the free behaviour through perturbative expansions in the coupling. When the latter is large, such expansions become meaningless, and one needs to know how the full theory behaves. In QCD, at low momenta quarks and gluons interact so strongly that they cannot exist as quasi-free objects any more, but are doomed to live confined to form colourless objects, the hadrons.

We are now in a position to understand qualitatively how jets are formed. Let us consider a quark produced in a high-energy process. The fact that it is ripped off the vacuum results in a huge instantaneous acceleration, similar to that experienced by fast electrons colliding on a target. Similarly to how those electrons emit bremsstrahlung photons, e.g. x-rays, an accelerated quark radiates gluons. In fact, at high energy the QCD coupling is small, and, if not for the colour matrices t^a and the fact that gluons are interacting, QCD closely resembles classical electromagnetism. As in classical electrodynamics, gluons are radiated preferably with small energies, and collinear to their emitter, with a probability density

$$dP(E, \theta) \sim \alpha_s(E\theta)\frac{dE}{E}\frac{d\theta^2}{\theta^2}, \tag{3.2}$$

where E is the energy of the emitted gluon and θ its angle relative to the emitting quark. Note that the coupling α_s is to be evaluated at the scale $E\theta$, the transverse momentum of the radiated gluon with respect to the emitting quark. The most likely configurations are those in which gluons are emitted at subsequently decreasing angles. This gives an ensemble of highly collimated partons. This 'branching' process changes when the angle and/or the energy of the last emitted parton are so small that $\alpha_s(E\theta)$ becomes large. At this point, the produced quarks and gluons feel very strong interactions with the neighbouring partons, and hadrons are formed. We do not know how hadronisation actually takes place, and we have to rely on models, whose assumptions are regularly validated against data. It happens that models in which only partons whose momenta are of similar size form hadrons provide the best description of collider data. We can then safely assume that, when the QCD coupling becomes large, neighbouring partons gather together to form hadrons, without a significant reshuffling of energy and momenta with respect to the parent partons. This is qualitatively why a highly energetic quark gives rise to a jet of highly collimated hadrons. Of course, the momenta of the hadrons in a jet will be related to that of the parent quark or gluon. For instance, the change in a jet transverse momentum $p_{\mathrm{t,jet}}$ induced by hadronisation is of the order of a small hadronisation scale ($\sim$1 GeV), so that the relative corrections ($\sim$1 GeV/$p_{\mathrm{t,jet}}$) are expected to decrease with increasing $p_{\mathrm{t,jet}}$. Furthermore, with increasing jet transverse momenta, given that the gluon emission probability in equation (3.2) is proportional to $\alpha_s(E\theta) \sim \alpha_s(p_{\mathrm{t,jet}}\theta)$, the angle of emitted gluons such that the QCD coupling becomes of order one and hadronisation takes place becomes smaller and smaller. This is why, at high transverse momenta, jets are so collimated that their momenta are very close to those of the individual quarks or gluons that have initiated them.

This qualitative picture of jet formation is confirmed by the excellent agreement between QCD theoretical predictions and experimental data. The rest of the chapter will thus be devoted to presenting the basics of QCD that are needed to understand jet physics and to describing the state-of-the art of QCD theoretical tools.

3.1 Fixed-order QCD calculations

The most natural way of performing theoretical calculations for jet observables is to use fixed-order QCD perturbation theory. This framework gives predictions that are truncated at a given order in the QCD coupling α_s. Before discussing the general features of fixed-order calculations, in particular their scope and their limitations, it is instructive to discuss first some elementary examples, which however contain the main features of such predictions.

3.1.1 Leading order (LO) calculations

Jet rates in e^+e^- annihilation. As an example, we consider e^+e^- collisions, in particular events containing hadrons in the final state. The majority of such events are characterised by the presence of two jets that are quasi back-to-back, i.e. point in opposite directions. They originate from the production of a quark and an antiquark, which undergo several branchings, and then hadronisation. A relevant fraction of events (around 10%–20%) is characterised by the presence of three energetic and well-separated jets. Here we wish to quantify this fraction using only fixed-order QCD perturbation theory. First, QCD tells us that these events are due to the emission of a hard gluon from the original quark–antiquark pair. Therefore, our first task is compute the probability for such an emission to occur, differential in suitable kinematic variables spanning the phase space of the emitted gluon. This is achieved by computing two contributions, represented pictorially by the two Feynman diagrams of figure 3.1. Their sum gives the quantum mechanical amplitude for the process under consideration, at the lowest order in QCD perturbation theory. This order is called the leading order (LO), or 'Born' level, a term inherited from scattering theory in non-relativistic quantum mechanics. The obtained amplitude has to be squared and suitably integrated over all possible values of the momenta in the final state, according to a Lorentz-invariant measure, which is known as multi-particle phase space. Denoting by p_q, $p_{\bar{q}}$ and p_g the momenta of the quark, anti-quark and gluon respectively, it is customary to introduce the kinematical variables

$$x_q = \frac{2(p_q q)}{Q^2}, \quad x_{\bar{q}} = \frac{2(p_{\bar{q}} q)}{Q^2}, \quad Q^2 \equiv q^2, \quad q \equiv p_q + p_{\bar{q}} + p_g, \tag{3.3}$$

with $Q = \sqrt{s}$ the centre-of-mass energy of the e^+e^- collision. From energy–momentum conservation, we obtain that both x_q and $x_{\bar{q}}$ have to be less than one. Also, $2 - x_q - x_{\bar{q}}$ is the fraction of energy (with respect to $Q/2$) carried by the gluon, and this has to be also less than one.

One can then compute the differential cross section in x_q and $x_{\bar{q}}$, normalised to the total cross section σ for the process $e^+e^- \to$ hadrons [1, 2]. Keeping only the first term in α_s, we obtain

$$\frac{1}{\sigma}\frac{d\sigma}{dx_q dx_{\bar{q}}} = C_F \frac{\alpha_s}{2\pi}\frac{x_q^2 + x_{\bar{q}}^2}{(1 - x_q)(1 - x_{\bar{q}})}\Theta(x_q + x_{\bar{q}} - 1), \tag{3.4}$$

where the Heaviside theta function accounts for the fact that x_q and $x_{\bar{q}}$ are restricted to the region $2 - x_q - x_{\bar{q}} < 1$. The factor $C_F = 4/3$ is a 'colour' factor that takes into account how the colour of the emitted gluon contributes to the differential

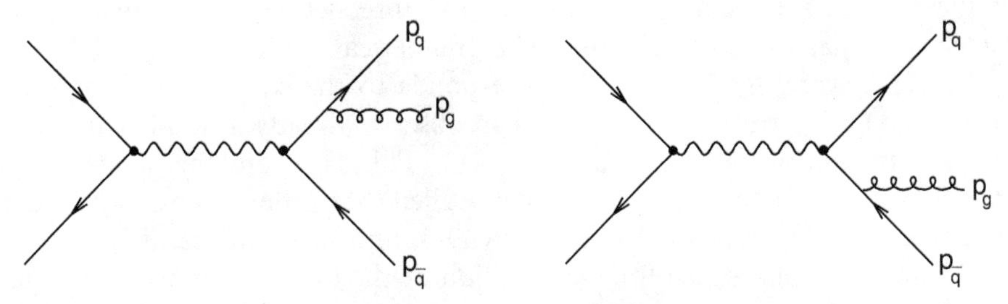

Figure 3.1. The two Feynman diagrams that give the quantum mechanical amplitude for $e^+e^- \to 3$ jets at the lowest order in QCD perturbation theory. The two lines with an arrow on the left-hand side of each diagram represent the incoming electron and positron, while the internal wavy line represents the sum of the contributions of all vector bosons (here a photon or a Z) that can mediate the interaction [1, 2].

cross-section in equation (3.4) (see section 3.2 for more details). From basic kinematic considerations based on momentum conservation, we obtain the relations

$$1 - x_q = \frac{(p_{\bar{q}} + p_g)^2}{Q^2}, \quad 1 - x_{\bar{q}} = \frac{(p_q + p_g)^2}{Q^2}, \quad x_q + x_{\bar{q}} - 1 = \frac{(p_q + p_{\bar{q}})^2}{Q^2}. \quad (3.5)$$

We observe that the differential cross-section in (3.4) gets singular when p_g becomes parallel (collinear) to p_q ($x_{\bar{q}} \to 1$) or to $p_{\bar{q}}$ ($x_q \to 1$), or when the energy of p_g vanishes, i.e. p_g becomes soft (both x_q and $x_{\bar{q}}$ tend to 1). Therefore, if we naively integrate the expression in equation (3.4) we obtain infinity. However, we can use equation (3.4) to compute the fraction of events in which three jets are observed, the so-called three-jet rate. To this end, we need a variable that discriminates between two- and three-jet events. For instance, we can cluster events into jets using the JADE algorithm described in section 2.1.2. We then introduce a jet resolution y_{cut}, and say we have three jets whenever $y_3 > y_{\text{cut}}$. In this case, $y_3 = \min[1 - x_q, 1 - x_{\bar{q}}, x_q + x_{\bar{q}} - 1]$, giving for the three-jet rate R_3 the following function of y_{cut}:

$$\begin{aligned} R_3(y_{\text{cut}}) = C_F \frac{\alpha_s}{2\pi} \int_0^1 dx_q \int_0^1 dx_{\bar{q}} \frac{x_q^2 + x_{\bar{q}}^2}{(1 - x_q)(1 - x_{\bar{q}})} \Theta(x_q + x_{\bar{q}} - 1) \\ \times \Theta(\min[1 - x_q, 1 - x_{\bar{q}}, x_q + x_{\bar{q}} - 1] - y_{\text{cut}}). \end{aligned} \quad (3.6)$$

The region of integration can be visualised in the plot of figure 3.2. There one sees that the regions in which the integrand is singular do not contribute to the three-jet rate. In other words, as explained in chapter 2, the three-jet rate is an infrared and collinear safe observable. We can then perform safely the integral in equation (3.6) and obtain

$$\begin{aligned} R_3(y_{\text{cut}}) = C_F \frac{\alpha_s}{2\pi} \Bigg[2\ln^2\left(\frac{y_{\text{cut}}}{1 - y_{\text{cut}}}\right) + (3 - 6y_{\text{cut}}) \ln\left(\frac{y_{\text{cut}}}{1 - 2y_{\text{cut}}}\right) + \frac{5}{2} - \frac{\pi^2}{3} \\ - 6y_{\text{cut}} - \frac{9y_{\text{cut}}^2}{2} + 4\text{Li}_2\left(\frac{y_{\text{cut}}}{1 - y_{\text{cut}}}\right) \Bigg] \Theta\left(\frac{1}{3} - y_{\text{cut}}\right). \end{aligned} \quad (3.7)$$

Equation (3.7) is a genuine prediction for the three-jet rate calculated using the language of quarks and gluons. The numerical value for LEP1 energy $Q = 91.2$ GeV, using $\alpha_s(Q) = 0.118$, corresponds to the solid curve in figure 3.3, labelled 'QCD LO'. From that plot we can observe already a number of features. First, the distribution vanishes for $y_{\text{cut}} = 1/3$. This is where the shaded area in figure 3.2 vanishes. It corresponds to the so-called 'Mercedes' events, where quark, antiquark and gluon have the same energy $Q/3$, and form angles of 120° between them. There, the phase space for the emission of the extra gluon from the quark–antiquark pair closes, and hence the three-jet rate is zero. With decreasing y_{cut}, there is more phase space available to the emitted gluon, so the three-jet rate increases. For smaller and smaller y_{cut} we probe the region where the gluon becomes soft or

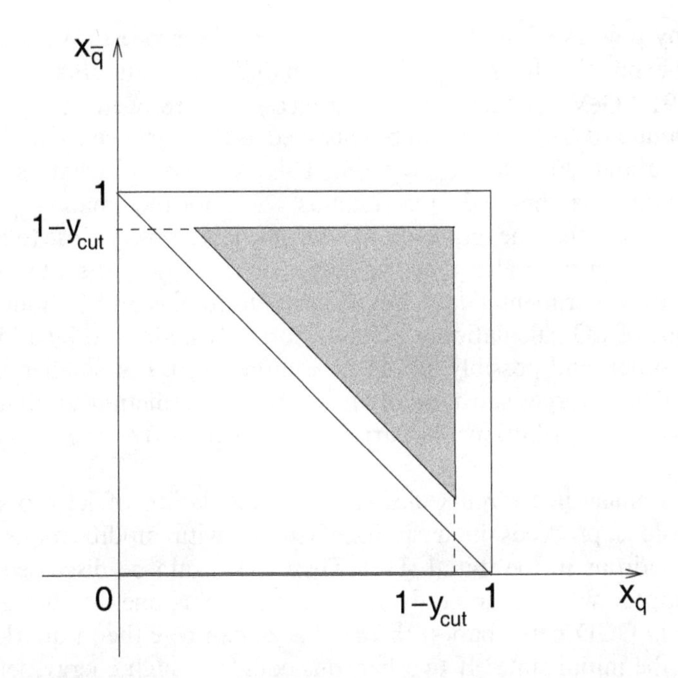

Figure 3.2. Allowed phase space for gluon emission from a $q\bar{q}$ pair in e^+e^- annihilation. The shaded region corresponds to the three-jet selection cut of equation (3.6).

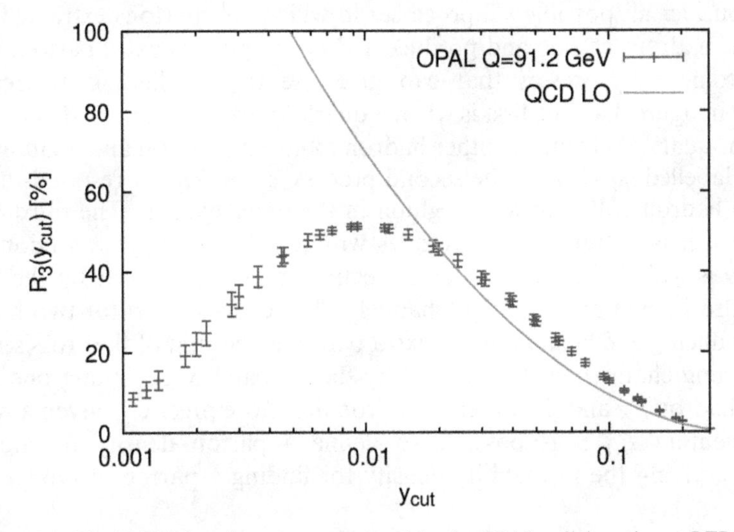

Figure 3.3. The JADE three-jet rate, as measured by the OPAL collaboration at LEP1 [3].

collinear to either the quark or the antiquark. In these regions, the squared amplitude, and hence the three-jet rate, diverges. We know that, in reality, jets are made of hadrons. However, as discussed in chapter 2, infrared and collinear (IRC) safe observables have the property that they can be safely compute at parton level, and the difference between parton and hadron level amounts to corrections suppressed by powers of the typical hard scale of the process. We can then try to compare the expression for $R_3(y_{\rm cut})$ in equation (3.7) with data obtained at the LEP1 energy $Q = 91.2$ GeV. For instance, using $\alpha_s(Q) = 0.118$, we can evaluate that for $y_{\rm cut} = 0.1$, around 10% of events will be classified as three-jet events, and this number increases to about 20% for $y_{\rm cut} = 0.05$. This is close to what is observed in experimental data (see figure 3.3), as long as $y_{\rm cut}$ is not too small ($y_{\rm cut} \gtrsim 0.01$), and confirms the picture that the momenta of hadronic jets are very close to those of their parent partons. We note also that, for large values of $y_{\rm cut}$, the LO curve follows qualitatively the experimental data, but its normalisation is off by about 20%. This is a general issue of LO calculations, which is normally addressed by adding the next perturbative order, and possibly more (see section 3.1.2). For smaller values of $y_{\rm cut}$, the shape of data diverges substantially from the LO prediction. In that region, one needs to consider contributions that arise at all perturbative orders, as explained in section 3.2.1.

Jet cross sections in hadron collisions. The calculation of jet cross sections in hadron collisions proceeds in a similar fashion, with modifications due to the presence of hadrons in the initial state. These can again be discussed through an example. Suppose we wish to study Z production plus one additional jet. At the lowest order in QCD perturbation theory this jet can be either a quark or a gluon. What about the initial state? If two hadrons collide at high energy, jet production occurs when they break apart, and two partons, one extracted from each hadron, give rise to an elementary collision. Therefore, if our final state is Z plus one jet, we need to consider all possible subprocesses in which two partons extracted from the initial-state hadrons collide and produce a Z boson plus an extra parton. There are three partonic subprocesses that can give rise to this final state, represented pictorially in figure 3.4. The first is when a quark (q) from the one hadron annihilates with an antiquark ($\bar{q}$) from the other hadron to give a Z boson and a gluon (g). This process is labelled $q\bar{q} \to Zg$. The second process, $qg \to Zq$, corresponds to a quark in the one hadron colliding with a gluon in the other hadron. The third $\bar{q}g \to Z\bar{q}$, occurs when it is antiquark that collides with a gluon. The cross section for each subprocess is called the partonic cross section, and is identified by the incoming partons (also known as incoming 'channel'). The cross section for two hadrons h_A and h_B producing a Z boson plus an extra parton is the sum of the cross sections for each incoming channel, each weighted by the probability of finding one incoming parton in hadron h_A and the other in hadron h_B. More precisely, given a hadron h_A with momentum P_A it is possible to define a parton density function (PDF) $f_{a/A}(x_a, \mu_F)$, giving the probability density for finding a parton of type $a = q, \bar{q}, g$

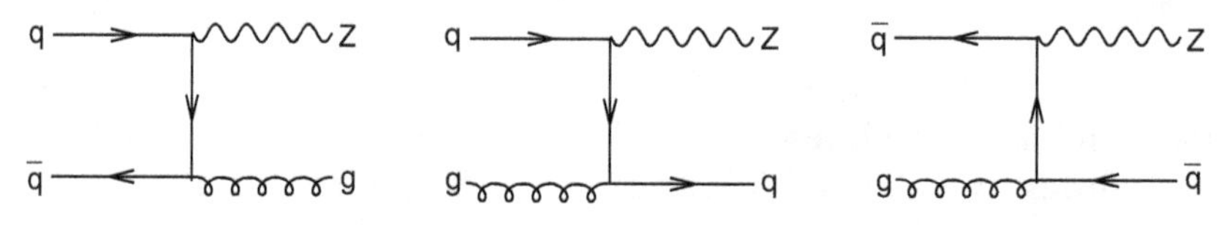

Figure 3.4. Representative Feynman diagrams for the three partonic subprocesses contributing to Z production plus one jet.

with momentum $p_a = x_a P_A$ inside hadron h_A [1, 2]. The scale μ_F, called factorisation scale, is an arbitrary energy scale on which each PDF depends, and represents the fact that the cross section we consider is insensitive to all partons with a transverse momentum (with respect to the beam) below μ_F. If we denote symbolically by $d\sigma_{ab\to Zc}[p_a, p_b]$ the cross section for the partonic subprocess $ab \to Zc$, fully differential in the momenta of the final-state particles, the hadronic cross section for Z plus one jet, at the lowest order in QCD perturbation theory, is given by

$$
\begin{aligned}
d\sigma_{h_A h_B \to Z+\text{jet}} = \int_0^1 dx_a \int_0^1 dx_b \\
\times \Bigg\{ \sum_q \Big(& f_{q/A}(x_a, \mu_F) f_{\bar{q}/B}(x_b, \mu_F)\, d\sigma_{q\bar{q}\to Zg}[x_a P_A, x_b P_b] \\
& + f_{\bar{q}/A}(x_a, \mu_F) f_{q/B}(x_b, \mu_F)\, d\sigma_{q\bar{q}\to Zg}[x_b P_B, x_a P_A] \Big) \\
+ \sum_q \Big(& f_{q/A}(x_a, \mu_F) f_{g/B}(x_b, \mu_F)\, d\sigma_{qg\to Zq}[x_a P_A, x_b P_B] \\
& + f_{g/A}(x_a, \mu_F) f_{q/B}(x_b, \mu_F)\, d\sigma_{qg\to Zq}[x_b P_B, x_a P_A] \Big) \\
+ \sum_{\bar{q}} \Big(& f_{\bar{q}/A}(x_a, \mu_F) f_{g/B}(x_b, \mu_F)\, d\sigma_{\bar{q}g\to Z\bar{q}}[x_a P_A, x_b P_b] \\
& + f_{g/A}(x_a, \mu_F) f_{\bar{q}/B}(x_b, \mu_F)\, d\sigma_{\bar{q}g\to Z\bar{q}}[x_b P_B, x_a P_A] \Big) \Bigg\}.
\end{aligned}
\tag{3.8}
$$

In the above expression, the sum over q extends to all the quarks that can be found in the proton at a scale below μ_F. For instance, if μ_F is of the order of some tens of GeV, $q = u, d, c, s, b$. The notation above is quite cumbersome, and not generally used in collider physics. When considering the subprocess $ab \to Zc$, it is understood that parton a (q or g) is extracted from hadron h_A, parton b (again q or g) is extracted from hadron h_B. The quark index q runs over both quarks and antiquarks, and the additional index $\bar{q}$ is introduced to denote the antiparticle of q (be it anti-quark or quark). If needed, an additional index q' denotes a quark with flavour strictly different from q. With this notation, we have three subprocesses, $q\bar{q} \to Zg$, $qg \to Zq$, $gq \to Zq$. With this simplified notation one obtains

$$
\begin{aligned}
d\sigma_{h_A h_B \to Z+\text{jet}} = \int_0^1 dx_a \int_0^1 dx_b \\
\times \Bigg\{ & \sum_q f_{q/A}(x_a, \mu_F) f_{\bar{q}/B}(x_b, \mu_F)\, d\sigma_{q\bar{q}\to Zg}[x_a P_A, x_b P_b] \\
& + \sum_q f_{q/A}(x_a, \mu_F) f_{g/B}(x_b, \mu_F)\, d\sigma_{qg\to Zg}[x_a P_A, x_b P_B] \\
& + \sum_q f_{g/A}(x_a, \mu_F) f_{q/B}(x_b, \mu_F)\, d\sigma_{gq\to Zq}[x_a P_A, x_b P_b] \Bigg\},
\end{aligned}
\tag{3.9}
$$

where now the index q runs on both quark and anti-quark flavours. In hadronic collisions, jet cross sections are usually presented as differential in the transverse momentum of the jet with respect to the beam axis $p_{\mathrm{t,jet}}$, and of the rapidity of the jet y_{jet}. If we wish to compute the hadronic Z+1jet cross section, differential in $p_{\mathrm{t,jet}}$ and y_{jet}, we need to compute $d\sigma_{ab\to Zc}[p_a, p_b]/(dp_{\mathrm{t,jet}}dy_{\mathrm{jet}})$ for each partonic subprocess contributing to our observable, and weigh each cross section with the appropriate parton densities. Typically, Z bosons are identified from their decay into a lepton–antilepton pair, observed within a selected 'fiducial' region of the detectors. Therefore, realistic cross sections must include also fiducial cuts in the decay products of the Z boson. Simulating such cuts requires numerical integrations, which are typically performed via Monte-Carlo methods and implemented in many publicly available programs. As an example we computed the $p_{\mathrm{t,jet}}$ distribution with the program MCFM [4], and compare it to LHC data taken by the ATLAS collaboration [5] (figure 3.5). The central theoretical prediction is obtained by choosing $\mu_{\mathrm{R}} = \mu_{\mathrm{F}} = \sqrt{p_{\mathrm{t,\,jet}}^2 + M_Z^2}$, with μ_{R} the so-called 'renormalisation scale', i.e. the scale at which the strong coupling is to be evaluated. The band represents an estimate of the theoretical uncertainty, obtained by varying μ_{R} and μ_{F} independently by a factor of two around the chosen central value within the range $1/2 \leqslant \mu_{\mathrm{R}}/\mu_{\mathrm{F}} \leqslant 2$. In fact, such a variation produces predictions that differ by a quantity of order α_s^2 from the central one, and can give an idea of the size of missing higher orders. The PDFs have been chosen from the PDF4LHC15_nlo_mc set, which implements the recommendations of the PDF4LHC working group [6]. We note first that the lowest order QCD prediction is close in shape to data, but underestimates them over the whole range of values of $p_{\mathrm{t,jet}}$. The discrepancy, although not very big, points to the fact that we do need missing higher orders to accurately describe the leading jet transverse momentum distribution. Also, the arbitrariness in the choice of the

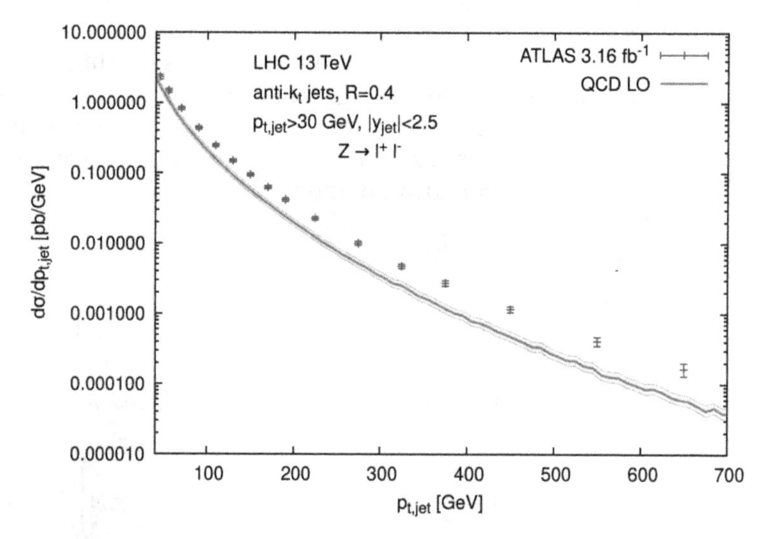

Figure 3.5. The distribution in the transverse momentum of the leading jet in Z plus one jet events, as measured by the ATLAS collaboration [5], compared to a LO QCD prediction obtained with the program MCFM [4].

central value of μ_R and μ_F is a ubiquitous characteristics of QCD calculations in hadron collisions, and calls again for the computation of higher orders to reduce the dependence on this choice. The main reason for this arbitrariness is that the centre of mass of each partonic collision is not known, so there is no unique hard scale that characterises a cross section in hadron collisions. One typically chooses a scale that is of the order of the transverse momentum of the jets involved, or of the masses of the heavy particles produced[1]. The scale we have chosen gives predictions that are close to experimental data, suggesting that higher order corrections are not too big, at least for this choice of scales. Despite their difficulties in describing data, lowest order calculations contain a lot of information on the physics of jet events. For instance, they give approximately the right shape for jet transverse momentum distributions, as well as jet angular correlations. Such observables, as in the case we have analysed, are in general not distorted in a significant way by higher order corrections, unless one looks into very specific configurations of final-state particles. This is why it is extremely important to be able to perform LO calculations in a general, fast and reliable way.

Tree-level techniques. LO calculations such as the ones outlined above are called 'tree-level', as opposed to loop calculations, whose representation in terms of Feynman diagrams involves particles forming loops[2]. The latter incorporate quantum corrections to tree-level amplitudes. Comparisons of tree-level calculations with experimental data has lead to an enormous amount of information on the nature of quarks and gluons. For instance, from angular correlations between pairs of jets in four-jet events in e^+e^- annihilation it was possible to confirm the existence of gluon self-interactions. The remarkable success of tree-level calculations pushed QCD practitioners to improve the technical tools for computing tree-level amplitudes with an increasing number of final-state particles (or 'legs'). On one hand, one needs an automated method to generate and evaluate Feynman diagrams. This can be done, for instance, with software packages such as QGRAF [7], FEYNARTS [8] or CALCHEP [9]. On the other hand, one needs techniques to integrate the resulting amplitude squared over the phase space of all particles in the final state. As done for the amplitude squared for the emission of a gluon from a $q\bar{q}$ pair, one usually parameterises this phase space in terms of a set of variables that assume values between zero and one, i.e. live inside a multi-dimensional hyper-cube. Actual experimental cuts on final-state particles restrict the range of integration to a multi-dimensional hyper-surface, whose boundary is generally so complicated that the integration cannot be performed with analytic methods, but only with Monte Carlo techniques. These procedures work as follows. One generates a sequence of

[1] There are no stringent rules in the choice of the central value for μ_R and μ_F in hadron collisions. A general understanding of the meaning of those scales provides sensible guidelines for this choice. This will be discussed to some extent in the next section, in the context of NLO calculations.

[2] Some calculations at the Born level can proceed through loops. For instance, Higgs production via gluon fusion proceeds via a loop of top quarks. Amplitudes for such processes are called 'loop induced', to distinguish them from tree-level ones.

random numbers $(x_1, \ldots, x_m)$ between zero and one. The sequence of generated random numbers $(x_1, \ldots, x_m)$ can be translated into a sequence of final-state momenta $(p_1, \ldots, p_n)$. At this point, one can check directly if the produced final-state momenta pass the required experimental cuts ($y_3 > y_{\rm cut}$ in our example in e^+e^- annihilation). If so, the product of the squared amplitude and of the phase-space Jacobian is the event weight, which is added to the histograms corresponding to the cross sections one wishes to compute. At the end of the Monte-Carlo procedure, the average over the number of generated events of the total weight in each histogram represents an estimator for the physical cross section.

For processes with many legs in the final state, it is essential to speed up this procedure as much as possible. On one hand, this is achieved by improving the efficiency of event generation. This is done by using adaptive techniques such as importance sampling (see e.g. [10]) or VEGAS [11], which aim at generating the most events where the integrand has the largest value, so as to make the Monte-Carlo procedure converge faster to the actual value of the integral we wish to compute. These techniques are implemented in all tree-level event generators used nowadays, like the aforementioned CALCHEP and MADGRAPH [12, 13]. The latter is a fully automated framework that generates tree-level amplitudes for all processes in the Standard Model, as well as some models of yet undiscovered new physics. MADGRAPH is able also to square amplitudes and feed them into a Monte Carlo event generator that provides histograms as requested by the user.

Another important aspect that requires consideration is the fact that the number of Feynman diagrams needed to compute an amplitude with a given number of legs grows factorially with the number of legs. This immediately creates a computational problem if we wish to describe jet-events at the LHC, where one expects to see events with a large number of jets. To give an idea, an amplitude for producing eight gluons in the final state results from the sum of more than one million of Feynman diagrams! A viable alternative to Feynman diagrams is helicity techniques [14, 15]. Gluons and quarks have two possible states of helicity, the component of the spin along the particle three-momentum. One decomposes the amplitude into all possible helicity states. As a first outcome, one gets that many of these amplitudes are related, which reduces the number of contributions to be computed. Second, some helicity amplitudes can vanish, which information cannot be easily obtained by just looking at individual Feynman diagrams. For instance, it can be shown that, if one considers gluons only, amplitudes in which all gluons have the same helicity vanish. The same happens when a single gluon has the opposite helicity with respect to the others. The first non-vanishing gluon amplitudes are the so-called maximally helicity-violating amplitudes, in which two gluons have opposite helicity with respect to all the others. Such amplitudes actually reduce to only one term [16], and more and more compact formulae are being found for all other helicity amplitudes (see e.g. [17–19]). Furthermore, one can exploit factorisation properties of amplitudes to devise recursion relations that link amplitudes with different number of particles [20–24]. Discussing these techniques is beyond the scope of this book. They constitute a very important topic in theoretical physics, for which the interest reader can find a review in [25].

3.1.2 Next-to-leading-order (NLO) calculations

In general, lowest order QCD calculations do not compare well with experimental data. In fact, corrections of relative order α_s are quite large, and are needed to achieve a good description of data, as well as control over renormalisation and factorisation scale uncertainties. We will then devote an entire section to explain how NLO calculations are performed. We will start with a fully worked-out example, the NLO calculation of the two-jet rate for the JADE algorithm. We will then discuss the tools that are available nowadays to perform such calculations.

3.1.2.1 Two-jet-rate in e^+e^- *annihilation*

Let us consider again jet production in e^+e^- annihilation. This time we wish to study the two-jet rate, the fraction of events that will be classified as two jets. We use again the JADE algorithm to cluster final-state particles into jets. At order α_s we have either zero or one extra gluon in the final state. As a consequence, according to the value of y_{cut}, we have two or three jets. In this situation, the two-jet rate $R_2(y_{\text{cut}})$ is just one minus the three-jet rate $R_3(y_{\text{cut}})$ in equation (3.6). Although this relation can be used to compute the two-jet rate, we wish to investigate how this result arises in fixed-order perturbation theory. Let us define $\sigma(y_{\text{cut}})$, the cross section for events with $y_3 < y_{\text{cut}}$, and σ_{tot} the total cross section for e^+e^- into hadrons[3]. Both cross sections admit an expansion in powers of α_s, evaluated at a renormalisation scale μ_R, as follows

$$\sigma = \sigma^{(0)} + \sigma^{(1)} + \sigma^{(2)} + \cdots, \tag{3.10}$$

where $\sigma^{(n)}$ is of relative order α_s with respect to $\sigma^{(n-1)}$. In this case, omitting for simplicity interactions mediated by a Z boson, we have

$$\sigma_{\text{tot}}^{(0)} = N_c \frac{4\pi\alpha}{3s} \sum_{\text{q}} Q_{\text{q}}^2, \qquad \sigma_{\text{tot}}^{(1)} = \frac{\alpha_s}{\pi} \sigma_{\text{tot}}^{(0)}. \tag{3.11}$$

Note that the expression in equation (3.11) is proportional to $N_c = 3$, the number of colours each quark can carry, as every colour contributes the same amount to the considered cross section. The sum in equation (3.11) extends to all quarks (each carrying electric charge $Q_{\text{q}}e$) whose mass is much smaller than $\sqrt{s}$, so that they can considered massless for the purpose of our calculation. For instance, at LEP1 energies, $\sqrt{s} = 91.2$ GeV, hence $q = u, d, c, s, b$. The two-jet rate can be expressed in terms of $\sigma(y_{\text{cut}})$ and σ_{tot} as follows

$$R_2(y_{\text{cut}}) = \frac{\sigma(y_{\text{cut}})}{\sigma_{\text{tot}}} = \frac{\sigma^{(0)}(y_{\text{cut}}) + \sigma^{(1)}(y_{\text{cut}}) + \sigma^{(2)}(y_{\text{cut}}) + \cdots}{\sigma_{\text{tot}}^{(0)} + \sigma_{\text{tot}}^{(1)} + \sigma_{\text{tot}}^{(2)} + \cdots}. \tag{3.12}$$

[3] As in section 3.1.1, we compute both cross sections using partons as final states, keeping in mind that they always turn into hadrons. Corrections induced by hadronisation will be discussed in section 3.3.

Strict fixed order for R_2 requires expanding the denominator, giving

$$R_2(y_{\text{cut}}) = \frac{\sigma^{(0)}(y_{\text{cut}}) + \sigma^{(1)}(y_{\text{cut}}) - \sigma_{\text{tot}}^{(1)} + \cdots}{\sigma_{\text{tot}}^{(0)}}. \tag{3.13}$$

At the lowest order, all events have two jets, irrespectively of the value of y_{cut}, hence $\sigma^{(0)}(y_{\text{cut}}) = \sigma_{\text{tot}}^{(0)}$. Therefore, at LO, $R_2(y_{\text{cut}}) = 1$ as expected. To compute $R_2(y_{\text{cut}})$ at the next perturbative order, the so-called NLO, we need to know $\sigma^{(1)}(y_{\text{cut}})$ and $\sigma_{\text{tot}}^{(1)}$. In fact, we have only to compute the difference between the two cross sections, which is simply given by $-\sigma_{\text{tot}}^{(0)} R_3(y_{\text{cut}})$, with $R_3(y_{\text{cut}})$ in equation (3.7). However, for the sake of illustration, we perform an explicit calculation of $\sigma^{(1)}(y_{\text{cut}})$. The NLO correction to the total cross section $\sigma_{\text{tot}}^{(1)}$ can be obtained for instance from $\sigma^{(1)}(y_{\text{cut}})$ by setting y_{cut} to its maximum value, i.e. $y_{\text{cut}} = 1/3$. The contribution $\sigma_{\text{R}}^{(1)}(y_{\text{cut}})$ to $\sigma^{(1)}(y_{\text{cut}})$ originating from the emission of a gluon from the original quark–antiquark pair is given by

$$\begin{aligned} \sigma_{\text{R}}^{(1)}(y_{\text{cut}}) = \sigma_{\text{tot}}^{(0)}\, C_{\text{F}} \frac{\alpha_{\text{s}}}{2\pi} \int_0^1 dx_q \int_0^1 dx_{\bar{q}} \frac{x_q^2 + x_{\bar{q}}^2}{(1 - x_q)(1 - x_{\bar{q}})}\, \Theta(x_q + x_{\bar{q}} - 1) \\ \times\, \Theta\big(y_{\text{cut}} - \min[1 - x_q, 1 - x_{\bar{q}}, x_q + x_{\bar{q}} - 1]\big). \end{aligned} \tag{3.14}$$

The integral above diverges when x_q and/or $x_{\bar{q}}$ equal one. This divergence occurs when the gluon becomes collinear to the quark ($x_{\bar{q}} \to 1$), to the antiquark ($x_q \to 1$) and/or when the gluon is soft ($x_q, x_{\bar{q}} \to 1$). As explained in the previous chapter, soft and collinear divergences cancel against quantum corrections arising when a gluon is emitted and then reabsorbed by the quark–antiquark pair. Such gluons are called 'virtual' because they do not appear in the final state, as opposed to 'real' particles, representing observable objects. We now discuss the explicit cancellation of soft and collinear divergences of the real cross section $\sigma_{\text{R}}^{(1)}(y_{\text{cut}})$ against virtual corrections, corresponding to the one-loop diagram in figure 3.6. Soft and collinear divergences are conveniently regularised by computing both real and virtual corrections in a number of space-time dimensions D that slightly differs from four, namely $4 - 2\epsilon$.

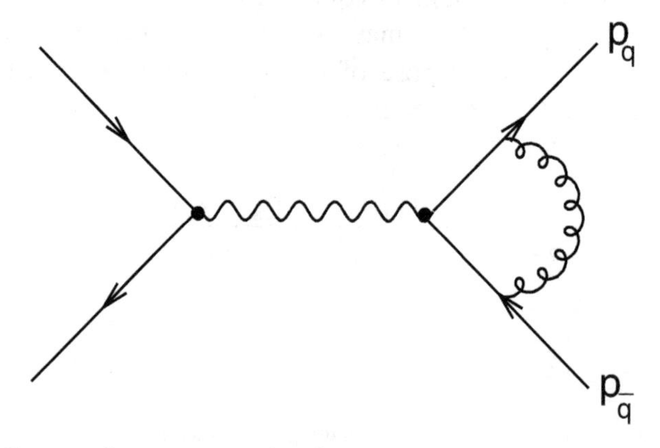

Figure 3.6. Feynman diagram representing virtual corrections to the two-jet rate at order α_{s}.

Virtual corrections are given as analytic functions of ϵ, with poles for $\epsilon \to 0$ corresponding to divergences. In particular, each soft or collinear divergence gives a $1/\epsilon$ pole. In our case, virtual corrections, arising from the interference between the virtual amplitude represented by the diagram in figure 3.6 and the tree-level amplitude, are given by $\sigma_{\rm V}^{(1)}(y_{\rm cut}) = V(\epsilon)\,\sigma_{\rm tot}^{(0)}$ with

$$V(\epsilon) = \frac{C_F\alpha_s}{2\pi}\left(\frac{4\pi\mu_R^2}{Q^2}\right)^{\epsilon}\frac{1}{\Gamma(1-\epsilon)}\left(-\frac{2}{\epsilon^2}-\frac{3}{\epsilon}+\pi^2-8+\mathcal{O}(\epsilon)\right). \tag{3.15}$$

In the above equation, the $1/\epsilon^2$ pole correspond to a soft and collinear divergence, whereas the $1/\epsilon$ pole to a collinear divergence. Besides infra-red and collinear divergences, virtual corrections might have additional $1/\epsilon$ poles due to ultra-violet divergences, arising when the momenta of the particles in the loops become very large. These divergences can be removed via a suitable redefinition of the parameters of the theory, for instance the coupling and the masses. This procedure, called 'renormalisation', ultimately results in the fact that the QCD coupling has to be defined in terms of some renormalisation scheme, and depends on an arbitrary renormalisation scale μ_R. Removing ultra-violet divergences corresponds to reabsorbing the contribution of high-momentum virtual particles in a redefinition of the parameters of a theory. This is why μ_R should be chosen of the order of the highest momentum scale in the process, in this case the centre-of-mass energy $\sqrt{s}$. Failure to do so results typically in large corrections involving logarithms of that scale and μ_R. We also remark that the dependence of cross sections on μ_R vanishes at all orders in perturbation theory. Fortunately, the virtual corrections in equation (3.15) are free from ultra-violet divergences, so no renormalisation is needed here. If it were the case, from now on we will assume that virtual corrections are renormalised, i.e. free from ultraviolet divergences.

In order to combine real and virtual corrections, we need also to compute the correction due to real emission in equation (3.14) in $4-2\epsilon$ dimensions. Expressing the result in terms of the variables x_q and $x_{\bar{q}}$ defined in equation (3.5) gives [1, 2]

$$\begin{aligned}\sigma_{\rm R}^{(1)}(y_{\rm cut}) = \sigma_{\rm tot}^{(0)}\,C_F\frac{\alpha_s}{2\pi}\left(\frac{4\pi\mu_R^2}{Q^2}\right)^{\epsilon}\frac{1}{\Gamma(1-\epsilon)}\int_0^1 dx_q\int_0^1 dx_{\bar{q}}\\ \times\frac{(x_q^2+x_{\bar{q}}^2)-\epsilon(2-x_q-x_{\bar{q}})^2}{(1-x_q)^{1+\epsilon}(1-x_{\bar{q}})^{1+\epsilon}(x_q+x_{\bar{q}}-1)^{\epsilon}}\,\Theta(x_q+x_{\bar{q}}-1)\\ \times\Theta(y_{\rm cut}-\min[1-x_q,1-x_{\bar{q}},x_q+x_{\bar{q}}-1]).\end{aligned} \tag{3.16}$$

Although it is possible to extract soft and collinear divergences from the above expression, it is not straightforward to separate the two, in that when the gluon becomes soft, both x_q and $x_{\bar{q}}$ tend to one. One can then extract the soft divergence by defining

$$z \equiv \frac{(p_q+p_{\bar{q}})^2}{Q^2}, \qquad \lambda \equiv \frac{(p_q+p_g)^2}{(1-z)Q^2}. \tag{3.17}$$

These definitions give very useful rescalings of the integration variables:

$$1 - x_{\bar{q}} = (1 - z)\lambda, \quad 1 - x_q = (1 - z)(1 - \lambda). \tag{3.18}$$

Before proceeding with the calculation, let us discuss the meaning of z and λ. From equation (3.18) we obtain

$$2 - x_q - x_{\bar{q}} = 1 - z = \frac{2(p_g \cdot q)}{Q^2}. \tag{3.19}$$

This implies that $1 - z$ is the fraction of the available energy $Q/2$ carried by the emitted gluon. Correspondingly, the invariant mass squared of the quark–antiquark pair, which was Q^2 before the emission of the gluon, reduces to zQ^2. Then, from equation (3.18), one sees that λ is an angular variable that is zero when the gluon is parallel to the quark, and one when it is parallel to the anti-quark.

We can now recast $\sigma_{\rm R}^{(1)}(y_{\rm cut})$ in terms of new variables z and λ, as follows

$$\begin{aligned}\sigma_{\rm R}^{(1)}(y_{\rm cut}) = \sigma_{\rm tot}^{(0)}\, C_{\rm F}\frac{\alpha_{\rm s}}{2\pi}\left(\frac{4\pi\mu_{\rm R}^2}{Q^2}\right)^{\epsilon}\frac{1}{\Gamma(1-\epsilon)}\int_0^1 \frac{dz\; z^{-\epsilon}}{(1-z)^{1+2\epsilon}}\int_0^1 \frac{d\lambda}{[\lambda(1-\lambda)]^{1+\epsilon}}\\ \times[1 + z^2 - 2\lambda(1-\lambda)(1-z)^2 - \epsilon(1-z)^2]\\ \times\Theta(y_{\rm cut} - \min[(1-z)\lambda,\,(1-z)(1-\lambda),\,z]).\end{aligned} \tag{3.20}$$

This expression has up to two poles in $1/\epsilon$, which can be extracted by using the following relation among distributions:

$$\frac{1}{x^{1+\epsilon}} = -\frac{1}{\epsilon}\delta(x) + \sum_{n=0}^{\infty}(-1)^n\,\epsilon^n\left(\frac{\ln^n x}{x}\right)_+, \tag{3.21}$$

where, given a test function $f(x)$, the 'plus' prescription is defined through the relation[4]

$$\int_0^1 dx\left(\frac{\ln^n x}{x}\right)_+ f(x) = \int_0^1 dx\left(\frac{\ln^n x}{x}\right)[f(x) - f(0)]. \tag{3.22}$$

[4] For a consistent application of the 'plus' prescription, it is crucial that the integral in equation (3.21) is between zero and one. If the integration over the x variable has an upper bound $x_0 < 1$, the plus prescription has to be interpreted as follows:

$$\begin{aligned}\int_0^{x_0} dx\left(\frac{\ln^n x}{x}\right)_+ f(x) &= \int_0^1 dx\left(\frac{\ln^n x}{x}\right)_+ f(x)\Theta(x_0 - x) = \int_0^1 dx\left(\frac{\ln^n x}{x}\right)[f(x)\Theta(x_0 - x) - f(0)]\\ &= \int_0^{x_0} dx\left(\frac{\ln^n x}{x}\right)[f(x) - f(0)] + \int_{x_0}^1 dx\left(\frac{\ln^n x}{x}\right)f(0)\end{aligned}$$

Before we apply the above identities to equation (3.20), it is useful to separate the collinear singularities for $\lambda \to 0$ and $\lambda \to 1$ using partial fractioning. This gives

$$\begin{aligned}\sigma_{\mathrm{R}}^{(1)}(y_{\mathrm{cut}}) = \sigma_{\mathrm{tot}}^{(0)}\, C_{\mathrm{F}}\frac{\alpha_{\mathrm{s}}}{2\pi}\left(\frac{4\pi\mu_{\mathrm{R}}^{2}}{Q^{2}}\right)^{\epsilon}\frac{1}{\Gamma(1-\epsilon)}\int_0^1 \frac{dz\, z^{-\epsilon}}{(1-z)^{1+2\epsilon}}\\ \int_0^1 \frac{d\lambda}{[\lambda(1-\lambda)]^{\epsilon}}\left(\frac{1}{\lambda}+\frac{1}{1-\lambda}\right)\\ \times[1+z^2-2\lambda(1-\lambda)(1-z)^2-\epsilon(1-z)^2]\\ \times\Theta(y_{\mathrm{cut}}-\min[(1-z)\lambda,(1-z)(1-\lambda),z]).\end{aligned} \tag{3.23}$$

The strategy is to expand every term in powers of ϵ, and keep only the poles and the constant parts, discarding all terms of order ϵ. While doing so, we observe that all terms containing poles in $1/\epsilon$ force the integrand to be evaluated in the soft or collinear limits, which implies that the observable constraint is always satisfied. Hence we can perform a number of integrations analytically and obtain[5]

$$\sigma_{\mathrm{R}}^{(1)}(y_{\mathrm{cut}}) = \sigma_{\mathrm{tot}}^{(0)}\, C_{\mathrm{F}}\frac{\alpha_{\mathrm{s}}}{2\pi}\left(\frac{4\pi\mu_{\mathrm{R}}^{2}}{Q^{2}}\right)^{\epsilon}\frac{1}{\Gamma(1-\epsilon)}\left(\frac{2}{\epsilon^2}+\frac{3}{\epsilon}+\frac{21}{2}-\pi^2+F(y_{\mathrm{cut}})+\mathcal{O}(\epsilon)\right), \tag{3.24}$$

where

$$\begin{aligned}F(y_{\mathrm{cut}}) \equiv \int_0^1 \frac{dz}{(1-z)_+}\int_0^1 d\lambda\left(\frac{1}{\lambda_+}+\frac{1}{(1-\lambda)_+}\right)\\ [1+z^2-2\lambda(1-\lambda)(1-z)^2]\\ \times\Theta(y_{\mathrm{cut}}-\min[(1-z)\lambda,(1-z)(1-\lambda),z])\end{aligned} \tag{3.25}$$

is by construction finite for all values of y_{cut}. We see that the poles in the real correction cancels exactly against those in the virtual correction. We can then add $\sigma_{\mathrm{V}}^{(1)}(y_{\mathrm{cut}})$ and $\sigma_{\mathrm{R}}^{(1)}(y_{\mathrm{cut}})$, take the limit $\epsilon \to 0$, and obtain

$$\sigma^{(1)}(y_{\mathrm{cut}}) = \sigma_{\mathrm{V}}^{(1)}(y_{\mathrm{cut}}) + \sigma_{\mathrm{R}}^{(1)}(y_{\mathrm{cut}}) = \sigma_{\mathrm{tot}}^{(0)} C_{\mathrm{F}}\frac{\alpha_{\mathrm{s}}}{2\pi}\left(\frac{5}{2}+F(y_{\mathrm{cut}})\right). \tag{3.26}$$

Note that the constant 5/2 in equation (3.26) is independent of the considered observable, although it depends on the way we have parametrised the phase space and the way we have regularised the singularities in the z, λ integrations. Let us consider now the expression for $F(y_{\mathrm{cut}})$ in equation (3.25). Expanding all plus distributions we get

[5] Note that, both in $\sigma_{\mathrm{V}}^{(1)}$ and $\sigma_{\mathrm{R}}^{(1)}$, the cross section $\sigma_{\mathrm{tot}}^{(0)}$ is understood to be calculated in $4-2\epsilon$ dimensions. However, since singularities cancel exactly between real and virtual corrections, so do the ϵ-dependent pieces of $\sigma_{\mathrm{tot}}^{(0)}$. This is why we have decided not to change notation with respect to the LO case.

$$\begin{aligned}F(y_{\text{cut}}) = &\int_0^1 \frac{dz}{(1-z)} \int_0^1 d\lambda \left(\frac{1}{\lambda} + \frac{1}{1-\lambda}\right)\\ &\times \left\{[1 + z^2 - 2\lambda(1-\lambda)(1-z)^2]\Theta(y_{\text{cut}} - \min[(1-z)\lambda, (1-z)(1-\lambda), z])\right.\\ &\left. - (1+z^2)\right\}.\end{aligned} \tag{3.27}$$

Before proceeding with the calculation of the expression in equation (3.27), we comment on its physical interpretation. The second line in that equation represents the contribution of the so-called 'real' events, each with a given value of z and λ giving some kinematic configuration of the $q\bar{q}g$ system, whose three-jet resolution y_3 is given by $\min[(1-z)\lambda, (1-z)(1-\lambda), z]$. The last line represents instead the contribution of the so-called 'counter-events', in which the emitted gluon is either collinear to the quark ($\lambda \to 0$), or to the antiquark ($\lambda \to 1$). There should also be a contribution in which the emitted gluon is soft ($z \to 1$) but not collinear, but its weight is zero in this particular case. In all these cases $y_3 = 0$, so that the counter-events live exactly where the real amplitude squared becomes singular, with a weight that is equal and opposite. Therefore, in actual calculations, soft an collinear singularities of real events do not cancel against virtual corrections, but against the appropriate counter-events. At this stage we can fully appreciate the implications of the IRC safety of a jet algorithm. If the algorithm were IRC unsafe, singular real events would not contribute the two-jet rate, and hence they would generate an infinite contribution that does not cancel with that of the corresponding counter-events.

After computing $F(y_{\text{cut}})$ explicitly we obtain $\sigma^{(1)}(y_{\text{cut}})$. Inserting the expression for $\sigma^{(1)}(y_{\text{cut}})$ in equation (3.13), we obtain the expected result

$$R_2(y_{\text{cut}}) = 1 - R_3(y_{\text{cut}}). \tag{3.28}$$

This completes the calculation of the two-jet rate for the JADE algorithm at NLO.

3.1.2.2 General features of NLO calculations

We now briefly discuss how the features discussed in the calculation of the two-jet rate are implemented in actual NLO calculations, focusing on a single aspect at a time.

Initial-state collinear singularities. In hadron collisions we meet an additional complication. In fact, $1/\epsilon$ poles of collinear origin do not cancel completely between real and virtual corrections. The reason is that a collinear emission from one of the initial-state partons takes a sizeable amount of energy from the emitting parton. Therefore, the total available centre-of-mass energy of the elementary partonic collision will be different for real and virtual corrections, causing a non-cancellation of collinear divergences. These divergences can, however, be reabsorbed in a redefinition of PDFs. In practice, one introduces extra universal counter-terms, proportional to $1/\epsilon$, which cancel against the surviving collinear singularities. One then repeats the same procedure as in e^+e^- annihilation, and, for an IRC safe observable, finds a finite results after the addition of real contributions, virtual corrections and the collinear counter-terms The result of this procedure is that cross sections depend on an unphysical factorisation scale μ_F. This corresponds to integrating inclusively over all collinear emissions up to the scale μ_F. Therefore, μ_F should be chosen of the order of the smallest momentum of resolved emissions.

Subtraction procedures. The procedure to subtract singularities described in the example of the two-jet rate is general enough to be applied to any fixed-order calculation. One maps the phase space for the emission of an extra parton onto a multi-dimensional hyper-cube, and for each independent variable performs an expansion in plus distributions as in equation (3.21). Although general, this procedure has the disadvantage that the parametrisation of the phase space is different for each process. It also requires analytical control over squared amplitudes, whose explicit expressions can be very complicated, especially for a large number of emitting partons. The most popular solution to this problem is to introduce universal counter-terms, which have the same singularities as real squared amplitudes, but are much simpler to handle analytically. In this case, real squared amplitudes can be computed in four dimensions, and their singularities are cancelled by the universal counter-terms The singularities of virtual corrections cancel against the integral of the counter-terms over their full phase space. The latter needs to be performed analytically in $4 - 2\epsilon$ dimensions. This last requirement restricts the possible choice of counter-terms In fact, only two procedures are widely used to perform NLO calculations, the first introduced by Frixione, Kunszt and Signer (FKS) [26], the other by Catani and Seymour [27]. Thanks to subtraction procedures, it is possible to construct NLO Monte-Carlo event generators that are fully differential in the momenta of final-state particles. These momenta can then be used to compute physical observables.

Phase-space slicing. Interestingly enough, the jet physics we have seen so far provides us with yet another method to eliminate divergences in real squared amplitudes. Let us consider the expression of the two-jet cross section for small values of y_{cut}:

$$\sigma^{(1)}(y_{\text{cut}}) \simeq \sigma_{\text{tot}}^{(0)} \frac{C_F \alpha_s}{2\pi} \left(-2 \ln^2 \frac{1}{y_{\text{cut}}} + 3 \ln \frac{1}{y_{\text{cut}}} + \frac{\pi^2}{3} - 1 \right). \tag{3.29}$$

We now generate events, and compute the value of y_3. If it is less than y_{cut} we consider the gluon as unresolved, i.e. the event will have the kinematics of a quark–antiquark event, and its weight will be given by equation (3.29). If $y_3 > y_{\text{cut}}$, the gluon is resolved, and the kinematics will be that of a $q\bar{q}g$ event. Suppose that we want to compute an observable in which the weight of resolved emissions with $y_3 \gtrsim y_{\text{cut}}$ contribute to the same histogram bin as unresolved emissions. This happens for instance in the case of σ_{tot}, the total cross section for $e^+e^- \to$ hadrons. The weight of resolved emissions will have up to two logarithms of y_{cut}, but these logarithms will cancel exactly those arising from unresolved emissions, displayed in equation (3.29), with a leftover that vanishes as y_{cut} becomes small. Therefore, for small-enough values of y_{cut}, any histogram in which resolved and unresolved emissions contribute in the same way to a physical observable will be independent of y_{cut} within the numerical precision. This procedure is called phase-space slicing, and can be performed with any IRC safe resolution variable that is different from zero when gluon emission occurs. Sometimes two resolution variables are used, one for the energy and one for the angle of the emitted gluon [28]. Due to the presence of the vanishing leftover, phase-space slicing is less used than subtraction methods. However, the method has been recently resurrected in the context of next-to-NLO (NNLO) calculations, as explained in section 3.1.3. It is intriguing how a basic knowledge of jet physics has lead already to a number of important applications.

Virtual corrections. The last ingredient of any NLO calculation is of course one-loop virtual corrections. This field experienced an important breakthrough with the development of the so-called 'unitarity' techniques, which solve the problem of computing one-loop amplitudes in a general and fast way. To explain how this works, we recall that any one-loop amplitude in QCD can be decomposed into a basis of known 'master' integrals, called boxes, triangles and bubbles because of their graphical representation in terms of Feynman diagrams for spin-0 particles. Such decomposition is usually performed for each Feynman diagram using the 'Passarino–Veltman' technique [29], consisting basically in the expansion of each integral into all possible Lorentz-covariant tensor structures. The main advance came from [30], where the coefficients of the decomposition in master integrals were related to physical amplitudes, which could be computed numerically using tree-level techniques, such as the helicity formalism discussed in the previous section. This led to a huge simplification in the calculation of virtual diagrams, and opened the way for efficient computation of QCD amplitudes with a large number of legs (see [31] for a review on the subject). These techniques made it possible, for instance, to compute for the first time the W production rate plus three jets at NLO [32]. Nowadays, unitarity techniques are implemented in a variety of computer programs that automatically generate virtual corrections. The most widely used are BLACKHAT [33], GoSAM [34] and HELAC [35]. Notably, BLACKHAT encodes W production up to five jets [36], Z production plus four jets [37], and four-jet production in hadron collisions [38]. In the current implementation, BLACKHAT is in fact responsible for the generation of virtual amplitudes, whereas subtracted real squared amplitudes with an extra parton are computed with the tree-level generator SHERPA [33]. GoSam is a publicly available package that automatically generates code for one-loop QCD amplitudes. It has been used for a number of multi-leg NLO calculation, including Higgs production plus three jets [39, 40]. In its current version, GoSam can be interfaced with both SHERPA and MADGRAPH. HELAC is a fully contained package for NLO calculations, including the efficient generation of real squared amplitudes using spinor-helicity amplitudes. One of its most important results is the NLO calculation of the cross section for $t\bar{t}b\bar{b}$ production [41], which is an important background to Higgs production, when the Higgs decays into a pair of W bosons. Unitarity techniques are also embedded in the package MADGRAPH_aMC@NLO, a fully automated framework that, combined with MADGRAPH, makes it possible, in principle, to compute NLO corrections to an arbitrary process [13]. This package is able also to interface NLO calculation with parton-shower event generators, as explained in section 3.2.1.

In parallel with the diffusion of unitarity techniques, numerical implementations of the 'traditional' Passarino–Veltman reduction technique are still being developed. Notably, $t\bar{t}b\bar{b}$ was also computed with Feynman diagram techniques, using automated reduction of one-loop amplitudes with the Passarino–Veltman method [42]. The automated procedure developed for those and similar calculations has been perfected further, and is now embedded in the publicly available library OpenLoops [43]. OpenLoops provides amplitudes up to one loop for a number of important multi-leg processes in e^+e^- annihilation and hadron collisions. OpenLoops is usually combined with SHERPA to include NLO corrections induced by real radiation (see [44] for a recent example).

After real and virtual corrections have been combined, NLO calculations are usually available in the form of Monte Carlo event generators, producing a finite number of partons in the final state. Besides the general purpose tools mentioned above, other popular NLO programs are EVENT2 [27] for e^+e^- annihilation, NLOJET++ [45], containing a selected number of processes in e^+e^- and hadron collisions, and MCFM [4], encoding many calculations for relevant processes at hadron colliders, such as W, Z, H and heavy-quark production.

In conclusion, NLO calculations constitute essentially a solved problem, the only issue being that of constructing efficient event generators that accommodate a large number of processes, and an arbitrary number of emitting legs. Practical limitations are the increasing number of Feynman diagrams that one has to compute, and the numerical stability of phase-space integrations. Such technical issues are still a subject of active research. Unfortunately, many of the techniques developed for NLO calculations do not work out of the box at higher orders, and new conceptual developments are needed, as will be argued in the next section.

3.1.3 Fixed-order calculations beyond NLO

There are several reasons why one would wish to compute jet cross sections beyond NLO, the most compelling one is clearly that of reducing the scale uncertainty of theoretical predictions, which for NLO calculations is still around 10%–20%. Pushing this accuracy to the percent level requires even higher-order corrections. Notably, an experimental accuracy below 10% has been already achieved by LHC experiments for many relevant cross sections at the end of the LHC second run, and will decrease even more with subsequent runs of the LHC. Furthermore, for many processes, including Higgs production plus jets, NLO corrections are as big as the LO contribution, suggesting a slow convergence of the QCD perturbative series. Improving such convergence requires in many cases imposing additional cuts on the final-state jets, for instance vetoing jets with transverse momentum above a given threshold. Precise predictions in the presence of tight kinematical cuts demand control on higher-order corrections. This gives some motivation behind pushing fixed-order calculations beyond NLO accuracy.

The first problem that arises at NNLO is the calculation of two-loop amplitudes. At the moment a basis of two-loop master integrals is not known. Therefore, one has to compute two-loop amplitudes on a case-by-case basis. These are known for all processes involving at most four external particles (see e.g. [46–49]), and for some processes involving five external particles [50–53].

The next problem is how to construct suitable counter-terms to eliminate soft and collinear divergences in real-emission squared amplitudes. Computing total cross sections is an easier task, as the cancellation of singularities can be performed before integrating over the available phase space. This is the reason why for specific total cross sections, like Higgs or vector boson production, we have calculations that extend up to N^3LO accuracy [54–56]. If we know the total cross section for a given process, inclusive in all coloured particles, but differential in the lowest order kinematics (e.g. Higgs inclusive cross section fully differential in the momentum of the Higgs boson), it is possible to construct local counter-terms by projecting each event into an underlying Born kinematics, and giving them a weight such that the inclusive integration over all coloured particles gives back the total cross section at the requested order. This is the strategy underlying the so-called 'projection to Born' method, which has been employed for the first time to generate fully exclusive events at NNLO in Higgs production via vector-boson fusion [57].

If the total cross section for a given process is not known, one has to devise local counter-terms. A general subtraction scheme with universal counter-terms has been proposed already for a long time [58], and all the counter-terms are currently available [59]. Although this method does provide local counter-terms, their integration, needed to cancel the singularities of virtual corrections, is technically challenging, and in the general case can be performed only numerically. Despite these difficulties, the method has been successfully implemented in the code ColorFullNNLO [60], with produces exclusive events for e^+e^- annihilation into three jets.

Another general subtraction procedure is the so-called antenna subtraction method [61]. The philosophy of the method is to construct counter-terms whose integral can be computed analytically, and cancel the singularities of virtual corrections. These counter-terms correspond to specific simplified cross sections, which can be computed fully analytically. Care must be taken that the integrands that give rise to these cross sections provide *local* counter-terms that cancel singularities of real squared amplitudes. The antenna method has been successfully applied to build the first-ever fully exclusive NNLO Monte Carlo event generator for $e^+e^- \to$ 3 jets [62]. It has been also applied to notable processes at hadron colliders, for instance Higgs [63] and vector-boson [64] plus jet, and dijet production [65]. All are implemented in the code NNLOJET.

Alternatively, a straightforward, although computationally involved, procedure, is to map the $4 - 2\epsilon$-dimension phase space for all real emissions onto a multi-dimensional hypercube, and perform an expansion in powers of ϵ. The construction of counter-terms proceeds as for NLO calculations if all singularities appear in a factorised form. Otherwise, it is possible to systematically split integrals in such a way that in each integrals all singularities are factorised. This method is known as sector decomposition, and can be generally applied to any integral in any number of dimensions [66, 67]. This is how the first exclusive NNLO Monte-Carlo event generators for Higgs [68] and vector boson [69] production in hadron collisions were constructed. An improvement on this line of thought consisted in a general phase-space parametrisation that makes it possible to perform the aforementioned ϵ expansion only on the singular limits of real-emission contributions [70]. This method has been used to obtain the first-ever NNLO generator for $t\bar{t}$ production [71]. It has also been applied to other processes at hadron colliders, for instance Higgs production plus one jet [72] and three-jet production [51].

Yet another strategy is to use a generalisation of the phase-space slicing technique described in section 3.1.2.2. Let us consider an IRC safe observable V, a function of all final-state parton momenta, that vanishes at the lowest considered order, and such that the cross section for $V < v$ is known at NNLO, up to corrections that vanish as v goes to zero. In this case, one can split the real-emission phase space so that configurations having $V > v_{\min}$ are resolved, hence give rise to partons in the final state, whereas configurations with $V < v_{\min}$ are unresolved, their weight being just the NNLO analytic cross section for $V < v_{\min}$. A Monte-Carlo event generator constructed in this way gives finite cross sections, and independent of $v_{\min}$ up to corrections that vanish for $v_{\min} \to 0$. One of these observables is the transverse momentum p_t of a Higgs or of a vector boson in hadron collisions. Knowledge of the most singular terms of the cross-section for $p_t < p_{t,\min}$ has been used to devise NNLO event generators for Higgs [73] and vector boson [74] production, and has been generalised to the production to an arbitrary colourless particle (also known as 'colour singlet') in hadron collisions, notably to Higgs plus vector boson [75, 76] and vector-boson pair production [77, 78]. NNLO calculations for the production of a colour singlet are implemented in the publicly available program MATRIX [79]. The same strategy has been also used to obtain NNLO predictions for $t\bar{t}$ production [80]. This generalised phase-space slicing method has been generalised to events with an arbitrary number of jets using the variable N-jettiness as a cut-off variable [81, 82]. Since numerical instabilities are worse at NNLO than at NLO accuracy, the cutoff $v_{\min}$ cannot be chosen too small. This has prompted the calculation of terms that are power suppressed in the resolution variable, see e.g. [83] for the case of N-jettiness.

The construction of NNLO Monte-Carlo event generators, fully differential in all final-state momenta, with more and more hard emitting partons, together with $\mathrm{N^3LO}$ predictions with lower multiplicities, is the frontier of QCD higher order calculations. Having as many of these calculations as possible is crucial for exploiting the full potential of the LHC, especially for the interpretation of deviations of data from Standard Model predictions.

3.2 QCD at all orders

Let us have a further look at the expression of the integrand in equation (3.20), and consider the limit $\lambda \to 0$. This limit corresponds to the gluon p_3 to be collinear to the quark p_1. In this limit, $\lambda \simeq z\theta^2$, where θ is the angle between the quark and the gluon. In that region, the integrand in question becomes

$$C_F \frac{\alpha_s}{2\pi} \frac{d\lambda}{\lambda} dz \frac{1+z^2}{1-z} \simeq \frac{d\theta^2}{\theta^2} dz \left[C_F \frac{\alpha_s}{2\pi} \frac{1+z^2}{1-z} \right]. \tag{3.30}$$

Let us comment on the physical meaning of the above equation. In the collinear limit $\lambda \to 0$, a quark of energy $E = \sqrt{s}/2$ emits a gluon with energy $(1 - z)E$, and its energy is reduced to zE. We say in this case that a quark has split into a gluon and a quark, and we denote this splitting as $q \to qg$. The quantity z is called the splitting fraction and the angle θ between the final-state quark and gluon is called the splitting angle. Equation (3.30) can be interpreted then as the probability for the splitting $q \to qg$ to occur with a splitting fraction between z and $z + dz$ and an angle between θ and $\theta + d\theta$. We see that the corresponding probability density is the product of a function of θ and a function of z. Also, the angular dependence is very simple, resulting only in a $1/\theta$ 'collinear' singularity. The dependence on z is more complicated, and features a $1/(1 - z)$ 'soft' singularity.

In general, it is possible to compute in QCD the probability density that a parton of type a (quark q, or gluon g) splits into two partons of type b and c, in quasi-collinear kinematics. The fact that it is possible to isolate an elementary subprocess in a quantum theory, in which probabilities are computed by squaring transition amplitudes, is a distinctive feature of the collinear limit of quantum field theory amplitudes. In general, let us consider a parton of type a with a given four-momentum p_a. By collinear splitting we mean the production of two partons p_b and p_c, quasi-parallel to p_a, as displayed in figure 3.7. All partons involved in the splitting can be considered to be quasi-massless, in the sense that their invariant masses are much less than their energies. In the collinear limit, the probability to produce partons p_b and p_c factorises into the product of the probability of producing p_a, times a universal factor, which can be interpreted as the elementary probability for the splitting $a \to b\,c$ [1, 2] with a splitting fraction between z and $z + dz$ and an angle between θ and $\theta + d\theta$. The splitting probabilities depend only on the type of parton involved (quarks or gluons), and can be expressed as

$$dP_{a\to b\,c}(z, \theta) = \frac{d\theta^2}{\theta^2} dz P_{a\to b\,c}(z) \frac{\alpha_s[z(1 - z)\theta E]}{2\pi}. \tag{3.31}$$

In the above expression, E is the energy of parton p_a, and zE the energy of parton p_b. By momentum conservation, the energy of parton p_c will be $(1 - z)E$. For the splitting to take place, p_a needs to have a positive invariant mass squared, which is approximately given by

$$p_a^2 = (p_b + p_c)^2 \simeq z(1 - z)\theta^2 E^2. \tag{3.32}$$

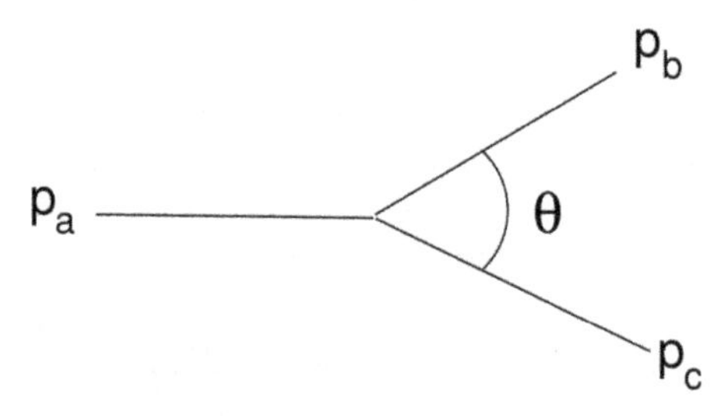

Figure 3.7. Pictorial representation of the splitting $a \to b\,c$.

Notice that the coupling has to be evaluated at the scale $z(1-z)\theta E$, which is of the order of the relative transverse momentum of partons p_b and p_c.[6] The splitting probability in equation (3.31) is large when the angle between p_b and p_c is small, i.e. p_b and p_c are quasi-collinear. In this situation $p_a^2 \ll E^2$. In the collinear limit $\theta \to 0$, the splitting probability diverges. This is precisely the collinear divergence of QCD squared amplitudes introduced in the previous chapter in the discussion about IRC safety. In the concrete example we have discussed in section 3.1.2, we find collinear singularities by taking the limits $\lambda \to 0$ and $\lambda \to 1$ in equation (3.20).

There are four basic splitting functions in QCD, $P_{q\to qg}(z)$, $P_{q\to gq}(z)$, $P_{g\to q\bar{q}}(z)$ and $P_{g\to gg}(z)$. Their expressions are (see e.g. [1])[7]

$$\begin{aligned}
P_{q\to qg}(z) &= C_F\frac{1+z^2}{1-z}, \quad P_{q\to gq}(z) = C_F\frac{1+(1-z)^2}{z}, \\
P_{g\to q\bar{q}}(z) &= T_F[z^2+(1-z)^2], \quad P_{g\to gg}(z) = C_A\left[\frac{z}{1-z}+\frac{1-z}{z}+z(1-z)\right].
\end{aligned} \tag{3.33}$$

Some comments are in order. First of all, the prefactors C_F, T_F, C_A are related to the gauge group of the theory. In particular, for a gauge theory based on the Lie group $SU(N_c)$, with N_c the number of colours, we have

$$\mathrm{Tr}(t^a t^b) = T_F\,\delta^{ab}, \qquad \sum_{a=1}^{N_c^2-1} t^a_{ik}t^a_{kj} = C_F\,\delta_{ij}, \qquad f^{acd}f^{bcd} = C_A\delta^{ab}. \tag{3.34}$$

The value of T_F is conventional, and sets the normalisation of the matrices t^a. We adopt the convention $T_F = 1/2$, giving

$$C_F = \frac{N_c^2-1}{2N_c}, \qquad C_A = N_c. \tag{3.35}$$

For $N_c = 3$ we have $C_F = 4/3$ and $C_A = 3$. Similar expressions hold in QED, with the only difference that $T_F = 1$, $C_F = 1$, and $C_A = 0$, so that the splitting of a photon into two photons cannot occur. Both QED and QCD share the fact that splitting functions can become singular when z is close to zero or one. This corresponds to the fact that one gauge boson (gluon or photon) has vanishingly small energy, i.e. becomes infinitely soft. This is precisely the soft divergence that has been discussed when the concept of IRC safety was introduced.

[6] To see this, let us consider the transverse momenta $\vec{p}_{tb}$ and $\vec{p}_{tc}$ with respect to the direction of a, with magnitudes $p_{tb} = zE\theta_b$, $p_{tc} = (1-z)E\theta_c$ and $\theta = \theta_b + \theta_c$. From momentum conservation $\vec{p}_{tb} + \vec{p}_{tc} = 0$, which implies that the magnitude of their relative transverse momentum $\vec{p}_{tb} - \vec{p}_{tc}$ is $2z(1-z)\theta E$.

[7] In the case of gluon splitting, in many situations one needs to take into account the fact that different gluon polarisations do not lead to the same splitting function, hence one gets an additional dependence on the azimuthal angle identifying the plane of the splitting. What we have presented so far are azimuthally averaged splitting functions, which are appropriate for all examples considered in this book.

The basics of jet physics can be understood by analysing the splitting probability in (3.31), and the splitting functions of (3.33). Suppose that an energetic parton p_a (quark or gluon) is produced in a high-energy collision, and a splitting $a \to bc$ occurs. We have three possible scenarios:

1. $\theta \ll 1$, z arbitrary: this is the basic process underlying jet formation. Parent parton p_a splits into two partons p_b and p_c, close in angle. These will subsequently branch to produce a cascade of quasi-collinear partons, which will turn into highly collimated bunches of hadrons. An IRC safe jet algorithm will likely cluster those hadrons as a single jet.
2. $\theta \sim 1$, $z(1-z) \ll 1$, and either parton b or parton c is a gluon: a soft gluon is emitted at large angles. This soft gluon is in general coherently emitted by all the hard partons in the event. Soft gluons at wide angles give rise to soft hadrons in the region between the jets, whose angular distribution reflects the colour connections of the hard emitters, the so-called 'colour flow' of the event (see [1] for the exact definition of what a colour connection is). The fate of such hadrons is either to be clustered with the nearest jet or, if too far away, give rise to soft jets.
3. $\theta \sim 1$, $z \sim 1$: a hard parton is produced, well separated from all other hard partons in the event. This will further split into a bunch of collinear partons, which will later hadronise and give rise to an extra jet.

QCD, in its perturbative formulation, is best suited to describe the third scenario. It costs a power of the coupling α_s to produce an extra jet. The requirement that all jets are energetic (hard) and well separated in angle ensures that jet production rates can be expressed as perturbative series in powers of α_s, which at high energies is indeed a small expansion parameter. In this region, the collinear approximation we have discussed here gives only a qualitative picture of the behaviour of hadronic jets and one has to perform fully fledged fixed-order perturbative calculations, as explained in section 3.1. Such calculations are the first that are attempted to describe jet observables, and play an extremely important role in our understanding of jet dynamics. This is why we have devoted the first section of this chapter to a general overview of the ideas underlying fixed-order QCD, and to an illustration of its most recent advances.

Despite the enormous success of fixed-order calculations in describing jet production rates, many interesting measurements involve looking inside the jets to unravel their structure. Such measurements normally require fixing a resolution parameter whose size is much less than the energy of the jets. The typical situation is the production of an object with a large mass, e.g. a Higgs boson, with a tight cut on the transverse momentum of accompanying jets. In such situations, a naive application of QCD perturbation theory leads in general to series that are poorly convergent. This is due to the appearance of large logarithms of two widely separated energy scales at all orders in perturbative expansions. In this regime, multiple soft and collinear branchings become relevant, and one can use the universality of splitting functions to obtain approximate predictions for interesting jet observables, either through analytical techniques (resummation) or through numerical simulations (the so-called 'parton-shower' event generators). The relevant feature of both approaches will be discussed in section 3.2.1.

The final aspect that needs to be discussed is the fact that jets are made of hadrons, not of quarks and gluons. It is legitimate therefore to ask the question: given a prediction for a jet observable in terms of quarks and gluons, will this be close to what we measure experimentally? In other words, what is the effect of phenomena, like hadronisation, which are beyond the domain of perturbative QCD? It is not possible to give a mathematically rigorous formulation of the problem. Nevertheless, it is possible to build realistic phenomenological models that, if successful in describing data, give some insight on the dynamics of intrinsically non-perturbative phenomena. Their implications for jet physics will be discussed in section 3.3.

3.2.1 Multi-parton branching

Let us consider again the e^+e^- two-jet cross section for $y_3 < y_{cut}$, in particular its expression for small values of y_{cut} in equation (3.29). This cross section contains logarithms of y_{cut} that diverge when y_{cut} becomes vanishingly small. This limit corresponds to having two extremely narrow jets, because events with any extra gluons will be classified as three-jet events and hence discarded. Such logarithms arise from the fact that one restricts the phase space available to real emissions, so that the cancellation of soft and collinear real and virtual contributions is not complete, but occurs only up to momentum scales of the order of $y_{cut}Q$, where $Q = \sqrt{s}$ represents the typical scale of the hard process at hand. Since both the energy and the angle of the extra gluon are logarithmically divergent, we obtain at most two logarithms of y_{cut} for each power of α_s. Such divergences for $y_{cut} \to 0$ are a signal of the breakdown of perturbation theory, in other words of the fact that considering a fixed number of extra emissions is not enough to obtain a physically sensible prediction for $\sigma(y_{cut})$ when y_{cut} is close to zero.

Notably, in the JADE algorithm, the three-jet resolution is closely related to the invariant mass of each jet. Hence, $y_{cut} \ll 1$ corresponds to the situation in which we require that the invariant mass of each jet be much less than its energy, which in two-jet events is of the order of the e^+e^- centre-of-mass energy Q. This is a very common situation in many experimental analyses. Suppose for instance that we are looking for a heavy particle (e.g. the Higgs) that decays hadronically. When the transverse momentum of the heavy particle is much larger than its mass, its decay products receive a huge boost in the heavy-particle direction, and are likely to clustered into the same jet. On top of the jets originated from the decay of the heavy particle, there will be background jets, whose invariant mass is dynamically produced through QCD radiation. It is therefore important to have theoretical tools that are able to tell us not only how likely it is that jets are produced, but also how their inner structure is determined by subsequent parton emissions. This is normally investigated by considering a jet resolution parameter (e.g. the jet invariant mass), and making it much smaller than the typical energy of the jets (e.g. the jet transverse momentum). Such analyses are then characterised by two scales, one being the typical energy of jets, and the other the small resolution parameter. Another situation characterised by the presence of two energy scales is cross sections with a jet-veto. There are situations in which one rejects events with jets with a transverse momentum above a given threshold. This is done for instance to suppress background from heavy coloured particles, such as top quarks, which will tend to produce many jets. The two scales in this case are the transverse momentum threshold, and the total invariant mass produced.

The physics of two-scale processes can again be understood from the simple example of the two-jet rate. When $y_{\mathrm{cut}} \ll 1$ we expect to find events with a quark and an antiquark without any accompanying gluons. But this is impossible, because a quark and an antiquark abruptly ripped off the vacuum will experience a huge instantaneous acceleration and, since they possess a colour charge, will radiate gluons. Hence we expect the two-jet rate to be exactly zero for $y_{\mathrm{cut}} = 0$. This behaviour cannot be obtained at any fixed order in perturbation theory, but only after resumming logarithmically enhanced contributions to all orders in perturbative expansions. A naïve resummation of the largest (leading) logarithms at all orders gives, for the JADE two-jet rate,

$$R_2(y_{\mathrm{cut}}) \simeq \mathrm{e}^{-\frac{\alpha_s C_F}{\pi} \ln^2 \frac{1}{y_{\mathrm{cut}}}} \to 0 \quad y_{\mathrm{cut}} \to 0. \tag{3.36}$$

The configurations that give rise to the exponential in the above equation contain an arbitrary number of soft gluons, collinear either to the quark or to the antiquark. Furthermore, the emissions collinear to each leg are strongly ordered in invariant mass, i.e. if $k_1, \ldots, k_n$ are collinear to the quark momentum p_q, we have $(k_1 p_q) \gg (k_2 p_q) \gg \cdots \gg (k_n p_q)$. Last but not least, the emission giving the largest invariant mass has to determine the value of y_3, i.e. $y_3 Q^2 \sim \max(\max_i\{2(k_i p_q)\}, \max_j\{2(k_j p_{\bar{q}})\})$, with $\{k_i\}$ collinear to the quark, and $\{k_j\}$ collinear to the antiquark $p_{\bar{q}}$. Unfortunately, for the JADE algorithm, not all configurations that are strongly ordered in invariant mass have this last property. Hence the resummation of the leading logarithms does not lead to an exponential, and equation (3.36) misses leading logarithms starting at order α_s^2 [84]. In practice, this reflects in the fact that the JADE algorithm can recombine soft gluons collinear to two different hard legs, as explained in section 2.1.2, and illustrated in figure 2.7. This is the main reason that brought QCD practitioners to devise 'exponentiating' jet algorithms, like the Durham or Cambridge algorithms. In both cases, resummation of the leading logarithms leads to an exponential similar to that in equation (3.36).

Exponentials like the one in equation (3.36) are called Sudakov form factors, and represent the probability of having no emissions above a given scale (represented for instance by a jet resolution variable). To understand how such exponentials emerge, we will abandon jet rates for a while, and discuss a simpler, and theoretically more transparent, example. In e^+e^- annihilation it is always possible to divide final state hadrons in two sets, called hemispheres, such that, at tree-level, the produced quark and antiquark belong to different hemispheres. Let us call a jet the set of all particles belonging to one hemisphere, and constrain the invariant mass (squared) of one hemisphere to be less than a given resolution $Q_0^2 \ll Q^2$. Suppose this jet contains the hard quark originated from the e^+e^- collision, and the quark emits a single collinear gluon. If z is the splitting fraction and θ the opening angle between the final-state quark and the emitted gluon, in the small-angle limit the invariant mass (squared) of the jet is $q^2 \sim z(1-z)\theta^2 Q^2$. Therefore, the probability $\Sigma(Q_0^2)$ that the mass of this jet is below Q_0^2 is just one minus the probability that its mass is above Q_0^2:

$$\begin{aligned}\Sigma(Q_0^2) &\simeq 1 - \int_0^1 \frac{d\theta^2}{\theta^2} \int_0^1 dz\, P_{q\to qg}(z) \frac{\alpha_s}{2\pi} \Theta\left[z(1-z)\theta^2 Q^2 - Q_0^2\right] \\ &\simeq 1 - \int_{Q_0^2}^{Q^2} \frac{dq^2}{q^2} \int_0^{1-q^2/Q^2} dz\, P_{q\to qg}(z) \frac{\alpha_s}{2\pi},\end{aligned} \tag{3.37}$$

where we have changed variable from θ^2 to $q^2 = z(1-z)\theta^2 Q^2$, and used the fact that the invariant mass of a jet cannot exceed Q^2. If in a hemisphere we have only collinear splittings with successively decreasing values of q^2, we have that $\Sigma(Q_0^2)$ is, in a first crude approximation, the probability of having no emissions with $q^2 > Q_0^2$. Let us split then the interval $[Q_0^2, Q^2]$ into n subintervals $[Q_{i-1}^2, Q_i^2]$, with $Q_n^2 = Q^2$. If these intervals are infinitesimally small, only one emission can have q^2 in any given interval, and therefore, the infinitesimal probability of not emitting anything in that interval is just

$$dP(Q_i^2, Q_{i-1}^2) = 1 - \int_{Q_{i-1}^2}^{Q_i^2} \frac{dq^2}{q^2} \int_0^{1-q^2/Q^2} dz\, P_{q\to qg}(z) \frac{\alpha_s}{2\pi}. \tag{3.38}$$

The total probability of having no emission with $Q_0^2 < q^2 < Q^2$ is the product of the elementary probabilities in (3.38):

$$\begin{aligned}\Delta_q(Q^2, Q_0^2) &= \lim_{n\to\infty} \prod_{i=1}^{n} dP(Q_i^2, Q_{i-1}^2) \\ &= \exp\left[-\int_{Q_0^2}^{Q^2} \frac{dq^2}{q^2} \int_0^{1-q^2/Q^2} dz\, P_{q\to qg}(z) \frac{\alpha_s}{2\pi}\right].\end{aligned} \tag{3.39}$$

The above expression gives a double logarithmic exponent such as the one in (3.36).

Not all configurations in which the mass of a jet is less than Q_0^2 are made up of strongly ordered emissions, so $\Sigma(Q_0^2)$ is only approximately equal to $\Delta_q(Q^2, Q_0^2)$. Corrections can be computed by introducing the jet invariant mass distribution $J_a(Q^2, k^2)$, the probability that a jet initiated by a parton of type $a = q, g$ has an invariant mass k^2, provided k^2 is less than Q^2. If we consider the mass of a jet as generated dynamically through subsequent collinear splittings, we observe that there are only two ways to produce a jet with invariant mass k^2. In fact, either no emissions occur, and the invariant mass of the jet is zero, or there is at least one splitting with opening angle θ and splitting fraction z, giving rise to two jets, each with an invariant mass below $q^2 = z(1-z)\theta^2 E^2$, with $E = Q/2$ the energy of the

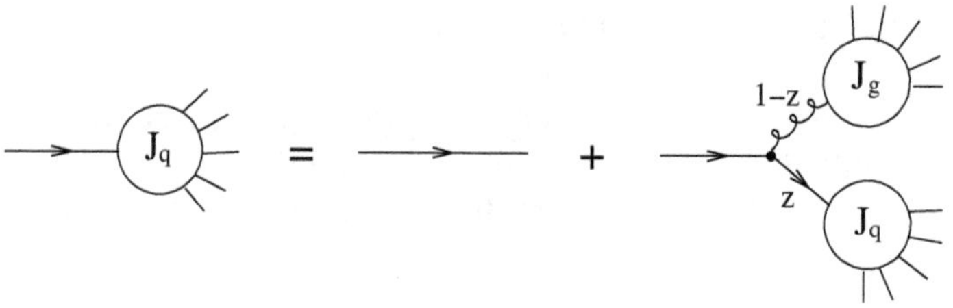

Figure 3.8. Pictorial representation of the evolution equation (3.40).

quark and antiquark produced in the hard e^+e^- collision. If a jet is initiated by a quark the various possibilities are illustrated in the figure 3.8. In formulae[8]

$$
\begin{aligned}
J_q(Q^2, k^2) = \Delta_q(Q^2, 0)\delta(k^2) + \int_0^{Q^2} \frac{dq^2}{q^2} \Delta_q(Q^2, q^2) \int_0^{1-q^2/Q^2} dz \, \frac{\alpha_s}{2\pi} P_{q\to qg}(z) \\
\times \int_0^{\infty} dq_1^2 \, J_q(q^2, q_1^2) \int_0^{\infty} dq_2^2 \, J_g(q^2, q_2^2) \, \delta\big(k^2 - q^2 - q_1^2 - q_2^2\big),
\end{aligned}
\tag{3.40}
$$

and an analogous equation holds for J_g, namely

$$
\begin{aligned}
J_g(Q^2, k^2) = \Delta_g(Q^2, 0)\delta(k^2) + \int_0^{Q^2} \frac{dq^2}{q^2} \Delta_g(Q^2, q^2) \int_0^{\infty} dq_1^2 \int_0^{\infty} dq_2^2 \\
\times \int_{q^2/Q^2}^{1-q^2/Q^2} dz \frac{\alpha_s}{2\pi} \Big[P_{g\to gg}(z) J_g(q^2, q_1^2) \, J_g(q^2, q_2^2) \\
+ n_f P_{g\to q\bar{q}}(z) J_q(q^2, q_1^2) \, J_q(q^2, q_2^2) \Big] \\
\times \delta\big(k^2 - q^2 - q_1^2 - q_2^2\big),
\end{aligned}
\tag{3.41}
$$

where n_f is the number of all massless quark flavours. The Sudakov form factor $\Delta_g(Q^2, q^2)$ can be inferred from the same argument leading to the expression for $\Delta_q(Q^2, q^2)$, and reads

$$
\Delta_g(Q^2, Q_0^2) = \exp\Bigg[-\int_{Q_0^2}^{Q^2} \frac{dq^2}{q^2} \int_{q^2/Q^2}^{1-q^2/Q^2} dz \, \frac{\alpha_s}{2\pi} \Big[P_{g\to gg}(z) + n_f P_{g\to q\bar{q}}(z) \Big] \Bigg].
\tag{3.42}
$$

In all the above equations we have assumed that there exists some well-defined procedure (e.g. dimensional regularisation) to regularise the otherwise vanishing Sudakov form factors $\Delta_q(Q^2, 0)$ and $\Delta_g(Q^2, 0)$. Also, from now on we will implicitly assume that all Sudakov form factors we will introduce are regularised. The fraction

[8] For historical reasons, q^2 is usually employed to denote the invariant mass (squared) in each splitting. The variable q should by no means be confused with the label q denoting a quark.

of events that have a quark jet with invariant mass less than Q_0^2 can be obtained by integrating equation (3.40) with respect to k^2 from 0 to Q_0^2. This gives

$$\Sigma(Q_0^2) = \Delta_q(Q^2, Q_0^2)\Bigg\{\Delta_q(Q_0^2, 0) + \int_0^{Q^2} \frac{dq^2}{q^2}\Delta_q(Q_0^2, q^2)\int_0^{1-q^2/Q^2} dz P_{q\to qg}(z)\frac{\alpha_s}{2\pi} \\ \times \int_0^{\infty} dq_1^2\, J_q(q^2, q_1^2)\int_0^{\infty} dq_2^2\, J_g(q^2, q_2^2)\,\Theta\big(Q_0^2 - q^2 - q_1^2 - q_2^2\big)\Bigg\}. \tag{3.43}$$

The expression in brackets has a finite expansion in powers of α_s, and, as will be clear later, contains less logarithms per power of α_s than the Sudakov form factor $\Delta_q(Q^2, Q_0^2)$, which represents therefore the dominant contribution to $\Sigma(Q_0^2)$.

Given the fact that distributions such as the one in the invariant mass of a jet are ubiquitous in all studies aimed at understanding the inner structure of jets, QCD practitioners have tried to describe such observables with the highest possible accuracy. This requires accounting for logarithmically enhanced contributions to all orders in QCD perturbation theory. There are two ways in which this theoretical programme is actually carried out, one is analytic resummations and the other is parton–shower event generators. In the following we will try to explain the philosophy underlying the two methods, highlighting advantages and limitations.

3.2.2 Analytic resummations

Let us consider again the case of the invariant mass of a hemisphere in e^+e^- annihilation (or of a jet, if we consider an event in hadronic collisions) as a relevant example of a jet observable. We have seen already that the probability $\Sigma(Q_0^2)$ that the jet mass is less than a given resolution Q_0^2 is roughly a Sudakov form factor $\Delta_q(Q^2, Q_0^2)$. The Sudakov form factor is an exponent

$$\Delta_q(Q^2, Q_0^2) \sim e^{-\alpha_s L^2}, \quad L = \ln\frac{Q}{Q_0}, \tag{3.44}$$

that contains at most one more power of the logarithm L than all powers of α_s. In fact, the Sudakov form factor aims at resumming all the so-called 'leading logarithmic' terms, those of the form $\alpha_s^n L^{n+1}$ in the *logarithm* of the mass distribution. Since most events lie in the region $\alpha_s L \sim 1$, leading logarithmic terms are not enough to constrain jet-observable distributions, but better logarithmic accuracy is needed. In the region $\alpha_s L \sim 1$, it is customary to reorganise the perturbative series for $\Sigma(Q_0^2)$ (as well as that for any other jet observable) as follows

$$\Sigma(Q_0^2) = e^{Lg_1(\alpha_s L)+g_2(\alpha_s L)+\alpha_s g_3(\alpha_s L)+\cdots}. \tag{3.45}$$

The function g_1 resums all leading logarithmic (LL) terms, of the form $\alpha_s^n L^{n+1}$, g_2 resums next-to-LL terms (NLL), $\alpha_s^n L^n$, g_3 resums next-to-NLL terms (NNLL), $\alpha_s^n L^{n-1}$, and so on. It is instructive to rewrite the perturbative series in the form

$$\Sigma(Q_0^2) = \mathrm{e}^{Lg_1(\alpha_s L)}[G_2(\alpha_s L) + \alpha_s G_3(\alpha_s L) + \cdots] \sim \mathrm{e}^L[1 + \alpha_s + \cdots]. \tag{3.46}$$

For $\alpha_s L \sim 1$ we have that all functions g_i are of order one, so that the only contribution that needs to exponentiate is that arising from g_1. The other contributions build up a new perturbative series, which starts from the NLL contribution which is of order 1, with higher logarithmic corrections suppressed by more and more powers of the QCD coupling. When discussing resummations, an important clarification is in order. While in fixed-order calculations moving from one order to the next is just a matter of computing more and more Feynman diagrams (which of course might be very complicated from a technical point of view), each logarithmic order already involves the summation of an infinite number of diagrams. Moving from one logarithmic order to the next require understanding which sets of diagrams, or, better yet, which physical effects are relevant at that order, as well as computing the corresponding contributions at the necessary accuracy.

Coherent branching. As an example, we discuss how NLL accuracy can be achieved for the jet mass distribution $\Sigma(Q_0^2)$, and other jet observables related to this quantity. A good starting point [85] is the evolution equation equation (3.40), which, as we have seen, is able to account for leading logarithms in the invariant mass distribution, but misses relevant NLL contributions arising from soft emissions. In fact, soft gluon emission does not admit a simple probabilistic interpretation, in that it depends crucially on the colour structure of all the energetic emitters. One crucial simplification however is the fact that a soft gluon emitted from a set of collinear partons sees only the total colour charge of the collinear ensemble rather than the individual colour charges (see e.g. [86], and references therein). Up to NLL accuracy, this 'coherence' property can be accounted for if subsequent collinear splittings are ordered in the splitting angle rather than in the invariant mass of the products of the splitting, as happens for the configuration shown in figure 3.9. Consequently, accounting for coherence in the jet-mass distribution requires changing the evolution variable from the invariant mass produced in each splitting to the relative angle θ of each splitting, as explained in detail in references [85, 87]. The resulting evolution equation reads

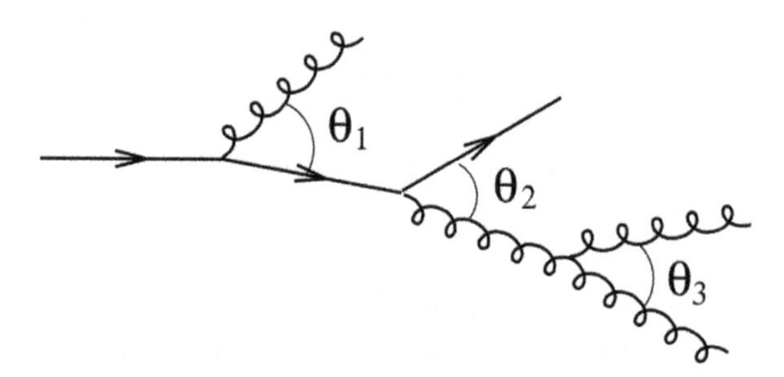

Figure 3.9. Three subsequent collinear splittings, with successively decreasing splitting angle $\theta_1 > \theta_2 > \theta_3$.

$$
\begin{aligned}
J_{\rm q}(Q^2, k^2) = \tilde{\Delta}_{\rm q}(Q^2, 0)\delta(k^2) + \int_0^{Q^2} \frac{d\tilde{q}^2}{\tilde{q}^2}\tilde{\Delta}_{\rm q}(Q^2, \tilde{q}^2) \\
\int_0^{1-\epsilon(\tilde{q}^2)} dz P_{q\to qg}(z)\frac{\alpha_{\rm s}^{\rm CMW}[z(1-z)\tilde{q}]}{2\pi} \\
\times \int_0^{\infty} dq_1^2\, J_{\rm q}(z^2\tilde{q}^2, q_1^2)\int_0^{\infty} dq_2^2\, J_{\rm g}((1-z)^2\tilde{q}^2, q_2^2) \\
\times \delta\left(k^2 - z(1-z)\tilde{q}^2 - \frac{q_1^2}{z} - \frac{q_2^2}{1-z}\right),
\end{aligned}
\tag{3.47}
$$

where $\tilde{q}^2 = \theta^2 E^2$ (where $E = Q/2$ for the first splitting) is a dimensionful variant of the splitting angle θ, and $\tilde{\Delta}_{\rm q}$ the Sudakov form factor corresponding to angular ordering. In the above equation, $\epsilon(\tilde{q}^2)$ is a cutoff that depends on the scheme used to regularize the z integration. This cutoff can be set to zero if the integrand is regular for $z \to 1$. To ensure NLL accuracy, the QCD coupling has to be evaluated in the physical CMW renormalisation scheme, related to the widely used $\overline{\rm MS}$ scheme by [85]

$$
\alpha_{\rm s}^{\rm CMW} = \alpha_{\rm s}^{\overline{\rm MS}}\left(1 + K\frac{\alpha_{\rm s}^{\overline{\rm MS}}}{2\pi}\right), \qquad K = C_{\rm A}\left(\frac{67}{18} - \frac{\pi^2}{6}\right) - \frac{10}{9}T_{\rm F}n_{\rm f}. \tag{3.48}
$$

The Sudakov form factor $\tilde{\Delta}_{\rm q}(Q^2, Q_0^2)$ reads

$$
\tilde{\Delta}_{\rm q}(Q^2, Q_0^2) = \exp\left[-\int_{Q_0^2}^{Q^2} \frac{d\tilde{q}^2}{\tilde{q}^2}\int_0^{1-\epsilon(\tilde{q}^2)} dz\, P_{q\to qg}(z)\frac{\alpha_{\rm s}^{\rm CMW}[z(1-z)\tilde{q}]}{2\pi}\right]. \tag{3.49}
$$

For the sake of analytic calculations, it is better to write the evolution equation for $J_{\rm q}(Q^2, k^2)$ in differential form. This is accomplished by rewriting equation (3.47) as follows:

$$
\begin{aligned}
\frac{J_{\rm q}(Q^2, k^2)}{\tilde{\Delta}_{\rm q}(Q^2, 0)} = \delta(k^2) + \int_0^{Q^2} \frac{d\tilde{q}^2}{\tilde{q}^2}\frac{\tilde{\Delta}_{\rm q}(Q^2, \tilde{q}^2)}{\tilde{\Delta}_{\rm q}(Q^2, 0)} \\
\int_0^{1-\epsilon(\tilde{q}^2)} dz P_{q\to qg}(z)\frac{\alpha_{\rm s}^{\rm CMW}[z(1-z)\tilde{q}]}{2\pi} \\
\times \int_0^{\infty} dq_1^2\, J_{\rm q}(z^2\tilde{q}^2, q_1^2)\int_0^{\infty} dq_2^2\, J_{\rm g}((1-z)^2\tilde{q}^2, q_2^2) \\
\times \delta\left(k^2 - z(1-z)\tilde{q}^2 - \frac{q_1^2}{z} - \frac{q_2^2}{1-z}\right).
\end{aligned}
\tag{3.50}
$$

By taking the derivative of equation (3.47) with respect to Q^2, and using the fact that

$$
Q^2\partial_{Q^2}\tilde{\Delta}_{\rm q}(Q^2, 0) = -\int_0^{1-\epsilon(Q^2)} dz\, P_{q\to qg}(z)\frac{\alpha_{\rm s}^{\rm CMW}[z(1-z)Q]}{2\pi}\tilde{\Delta}_{\rm q}(Q^2, 0), \tag{3.51}
$$

we can eliminate the Sudakov form factors completely, and obtain

$$Q^2\partial_{Q^2}J_q(Q^2, k^2) = \int_0^{1-\epsilon(Q^2)} dz\, P_{q\to qg}(z)\frac{\alpha_s^{\mathrm{CMW}}[z(1-z)Q]}{2\pi} \times\left[\int_0^\infty dq_1^2\, J_q(z^2Q^2, q_1^2)\int_0^\infty dq_2^2\, J_g((1-z)^2Q^2, q_2^2) \times\delta\left(k^2 - z(1-z)Q^2 - \frac{q_1^2}{z} - \frac{q_2^2}{1-z}\right) - J_q(Q^2, k^2)\right]. \tag{3.52}$$

This is a non-linear equation whose full solution is not needed to compute $J_q(Q^2, k^2)$ at NLL accuracy. In fact, coherence forces the angle of a subsequent branching of the gluon $q_2^2/(1-z)^2$ to be less than Q^2. Therefore, q_2^2 gives a non-negligible contribution to the delta function in equation (3.52) only for $q_2^2 \simeq (1-z)^2Q^2$, which gives a correction of relative order α_s, and hence NNLL. Neglecting this correction, one can integrate $J_g((1-z)^2Q^2, q_2^2)$ freely over q_2^2. This integration gives 1, because it corresponds to a total probability, so that in the end we obtain a linear equation. Furthermore, since $P_{q\to qg}(z)$ is singular at $z = 1$, one can set $z \to 1$ in all smooth functions of z, up to NNLL corrections. One then obtains the linear equation

$$Q^2\partial_{Q^2}J_q(Q^2, k^2) = \int_0^1 dz\, P_{q\to qg}(z)\frac{\alpha_s^{\mathrm{CMW}}[(1-z)\tilde{q}]}{2\pi} \times\left[J_q(Q^2, k^2-(1-z)Q^2) - J_q(Q^2, k^2)\right], \tag{3.53}$$

which can be solved analytically via an integral transform. Note that, in equation (3.53), the upper bound of the z integration has been set to 1, as the integrand is finite for $z \to 1$. Linear equations such as (3.53) can be written in an even more compact way by introducing the splitting kernel

$$P_{qq}(z) = C_F\left(\frac{1+z^2}{1-z}\right)_+ = C_F\left[\frac{1+z^2}{(1-z)_+} - \frac{3}{2}\delta(1-z)\right], \tag{3.54}$$

with the plus distribution $1/(1-z)_+$ defined in equation (3.22). This gives

$$Q^2\partial_{Q^2}J_q(Q^2, k^2) = \int_0^1 dz\, P_{qq}(z)\frac{\alpha_s^{\mathrm{CMW}}[(1-z)\tilde{q}]}{2\pi} J_q(Q^2, k^2-(1-z)Q^2). \tag{3.55}$$

Splitting kernels, obtained by combining real emission corrections and virtual corrections arising from Sudakov form factors, are the basic building blocks of the linear DGLAP equations that determine the dependence of PDFs on the factorisation scale (see e.g. [1]). Whether equation (3.53) accounts for all NLL contributions to e^+e^- physical observables that we can construct from the jet-mass distribution depends on the specific observable we consider. A way to probe the invariant mass of jets is through the two variables thrust and heavy-jet mass. In two-jet events, one minus the thrust is approximately the sum of the (squared) invariant masses of the two hemispheres (normalised to Q^2), whereas the heavy-jet mass is by definition the (squared) invariant mass of the heavier hemisphere (again normalised to Q^2). Both are examples of event-shape variables, because their value is correlated to the shape of the energy-momentum flow of hadronic events. Let us consider the cumulative distributions $\Sigma(\tau)$, the fraction of events for which one minus the thrust is less than τ. When τ is close to zero, events contain two pencil-like jets, whereas for large τ events are more symmetric, quasi-spherical. In the two-jet limit, i.e. if τ is close to zero, $\Sigma(\tau)$ is related to $J_q(Q^2, k^2)$ as follows

$$\Sigma(\tau) = \int_0^{\tau Q^2} dq_1^2\, J_q(Q^2, q_1^2) \int_0^{\tau Q^2} dq_2^2\, J_q(Q^2, q_2^2)\, \Theta(\tau Q^2 - q_1^2 - q_2^2), \tag{3.56}$$

with q_1^2 and q_2^2 the (squared) invariant masses of the two hemispheres. Similarly, for $\Sigma(\rho_H)$, the fraction of events for which the heavy-jet mass is less than ρ_H, we have

$$\Sigma(\rho_H) = \int_0^{\rho_H Q^2} dq_1^2\, J_q(Q^2, q_1^2) \int_0^{\rho_H Q^2} dq_2^2\, J_q(Q^2, q_2^2)\, \Theta\left(\rho_H Q^2 - \max\{q_1^2, q_2^2\}\right). \tag{3.57}$$

Note that, in writing equations (3.56) and (3.57) we have completely neglected correlations between the two hemispheres. In principle, they do appear, but for these observables only at subleading logarithmic accuracy. If we consider instead the cumulative distribution in the invariant mass of a single hemisphere, equation (3.53) is not enough to achieve full NLL accuracy. This is because, configurations in which soft emissions falling into the opposite hemisphere coherently emit a softer gluon that contributes to the mass of the observed hemisphere give a NLL contribution [88]. Such NLL effects are known as non-global logarithms, because they appear whenever the region in which measurements are performed is restricted. In fact, for the single-jet mass, gluons emitted in the unobserved region have energies that are logarithmically distributed between the hard scale of the process Q and jet resolution Q_0, giving rise to single-logarithmic contributions to all orders. For the thrust and heavy-jet mass, such configurations cannot occur because the momenta of all soft emissions are bound by the jet resolution, which is small and not of order Q. In practice, for the single-jet mass distribution $\Sigma(\rho)$, the probability that the (squared) invariant mass of one jet is less than ρQ^2, we get

$$\Sigma(\rho) = \int_0^{\rho Q^2} dq^2 J_q(Q^2, q^2) \int_0^{\rho Q} dk\, S_{q\bar{q}}(Q, k)\, \Theta(\rho Q^2 - q^2 - kQ), \tag{3.58}$$

where $S_{q\bar{q}}(Q, k)$ embodies the contribution of non-global logarithms. An evolution equation like equation (3.53) has been written for the scalar sum of transverse momenta within a jet, which led to the NLL resummation of the total and wide-jet broadenings, two relevant event shapes in e^+e^- annihilation, giving an idea of the width of jets [89, 90]. The distribution in the single-jet broadening is affected by non-global logarithms, and hence cannot be expressed in terms of the distribution in the scalar sum of transverse momenta within a jet.

Non-global logarithms. Non-global logarithms deserve a special mention. First of all, they are single logarithms, i.e. they give one logarithm for each power of α_s.

Their originate from soft emissions, which cannot be taken into account via a simple probabilistic picture. However, the probability of multiple soft emissions is known in the closed form in the limit of large number of colours (large-N_c) [86]. One can exploit this property to write a non-linear evolution equation whose solution provides the resummation of leading (single) non-global logarithms [91]. The equation has been solved first via a Monte-Carlo iteration [88, 92], and later on through other numerical techniques [93]. Currently, the LL evolution equation for non-global logarithms has been extended beyond the large-N_c limit [94], and recently to NLL accuracy [95–99], i.e. taking into account contributions that are suppressed by one power of α_s with respect to LL ones.

In hadron collisions, beyond the large-N_c limit, a further problem arises. In fact, virtual corrections connecting the two initial-state legs modify the colour of the initial state in such a way that non-global logarithms receive contribution from the collinear region. They then give double-logarithmic corrections, that are known as 'super-leading' logarithms [100]. Super-leading logarithms have been recently resummed using effective field theory methods [101].

Resumming non-global logarithms to high logarithmic accuracy is still a technically very challenging task. Therefore, to ensure that analytic resummations can be performed with the highest possible accuracy, one preferably considers observables free from non-global logarithms. This is what we will do in the rest of our discussion on resummation.

Soft-collinear effective theory. The limitation of the coherent branching formalism is that it is tied to a probabilistic picture of QCD collinear splitting, that might not be adequate to capture corrections beyond NLL accuracy. A more general analytic approach is based on Soft-Sollinear Effective Theory (SCET) [102]. In SCET one considers all Feynman diagrams that contribute to a given observable, and divides them according to how the loop momenta scale with respect to the physical scales of the problem. For instance, in the case of the thrust distribution the relevant scales are Q^2, τQ^2 and $\tau^2 Q^2$. It then possible to show that, in SCET, the thrust distribution for $\tau \ll 1$ can be written as [103]

$$\Sigma(\tau) \simeq H(Q^2, \mu) \int dq_1^2\, dq_2^2\, dk_s\, J(q_1^2, \mu) J(q_2^2, \mu) S(k_s, \mu) \\ \Theta\left(\tau Q^2 - q_1^2 - q_2^2 - k_s Q\right). \tag{3.59}$$

The hard function $H(Q^2, \mu)$ contains the contribution of all loop momenta of the order of the e^+e^- centre-of-mass energy Q. This implies that only virtual corrections can contribute to H, because a real emission with large momentum would give a jet with a large invariant mass, which cannot contribute to $\Sigma(\tau)$ in the two-jet limit. The *jet functions* $J(q_1^2, \mu)$ and $J(q_2^2, \mu)$ contain loop momenta which are energetic, but have a small invariant mass, of order $q_1^2 \sim q_2^2 \sim \tau Q^2$, as is typical for collinear, but energetic emissions. The *soft function* $S(k_s, \mu)$ contains loop momenta whose components are all small, and of order $k_s \sim \tau Q$. There are many other possible scalings of loop momenta, but they all lead to dimensionless integrals, which vanish when evaluated in dimensional regularisation. These functions, in particular the jet and soft functions contributing to the thrust, are all defined in terms of operators in an effective theory, even beyond perturbation theory, and have a well-defined expansion in powers of the QCD coupling. Furthermore, each function depends on an unphysical scale μ on which the thrust distribution $\Sigma(\tau)$ cannot depend. Taking the derivative with respect to μ gives an equation relating H, J and S, that leads to the emergence of the Sudakov form factor as the solution of a linear differential equation, rather than from probabilistic arguments. Also, each of the three functions satisfy a linear differential equation on its own, which is enough to fully constrain $\Sigma(\tau)$ at all logarithmic orders. SCET has been successfully applied to many final state observables in e^+e^- and hadron–hadron collisions, reaching an accuracy that is difficult to obtain with other approaches (see e.g. [104–107] for some of the most prominent achievements). This is also due to the fact that evolution equations for hard, soft and jet functions can be studied separately, and in some cases the results can be already found in the literature from unrelated studies. For instance, the hard function $H(Q^2, \mu)$ is just the magnitude squared of the on-shell quark form factor, a well-known object in quantum field theory. This has been computed up to four orders in α_s [108], and its functional form is known at all orders [109]. Also, the evolution equations for H, J and S all involve the so-called 'cusp anomalous dimension', which is another important object in quantum field theory, whose expansion in powers of the QCD coupling is known up to four orders [110]. Despite its successes, SCET has some limitations. First, the relevant expansion modes, and hence the needed effective theory, might differ from one observable to the next, and at the moment there exists no general criterion that associates an effective theory to a given observable. Second, the observable constraint, e.g. $\Theta\left(\tau Q^2 - q_1^2 - q_2^2 - k_s Q\right)$ has to be simple enough to be handled analytically in order to write factorised expressions ('factorisation theorems' in SCET terminology) such as the one in equation (3.59).

General resummation of recursive IRC safe observables. The main limitation posed by evolution equations is that physical observables need to be written in some factorised form. In fact, this is not the case for most final-state observables, especially for jet rates, which result from algorithmic procedures involving all final-state particles in a highly non-trivial way. However, the physics content of the coherent branching equations needed to achieve NLL resummation for the thrust and jet broadening distributions is that relevant QCD emissions at that accuracy are just soft and collinear gluons widely separated in angle. These gluons are independently emitted from the hard quark and antiquark produced in e^+e^- annihilation, and subsequent collinear splittings of the emitted gluons contribute beyond NLL accuracy. This corresponds to integrating away the secondary gluon mass distribution $J_g((1 - z)^2 Q^2, q_2^2)$ in equation (3.52).

Consider now a generic final-state observable $V(\{\tilde{p}\}, k_1, \ldots, k_n)$ in e^+e^- annihilation, where $\{\tilde{p}\}$ represent the primary quark and antiquark produced in the collision, and $k_1, \ldots, k_n$ are secondary partons. Suppose also that V vanishes when no emissions are present, i.e. $V(\{\tilde{p}\}) = 0$. If relevant emissions at NLL accuracy are the same as for the thrust and jet broadening, it is possible to write a general NLL resummation formula for $\Sigma(v)$, the fraction of events with $V(\{\tilde{p}\}, k_1, \dots, k_n) < v$. Consider soft gluons only, we obtain

$$\Sigma(v) \simeq e^{-\int [dk]M^2(k)} \sum_{n=0}^{\infty} \frac{1}{n!} \int \prod_{i=1}^{n} [dk_i]M^2(k_i)\, \Theta(v - V(\{\tilde{p}\}, k_1, \dots, k_n)), \quad (3.60)$$

where $M^2(k_i)$ is an effective emission probability for soft gluons widely separated in angle, and $[dk_i]$ a suitable Lorentz-invariant phase-space measure for gluon k_i. The product of independent emission probabilities corresponds to the contribution of real emissions, while virtual corrections are represented by the exponent $\exp\left[-\int [dk]M^2(k)\right]$, devised in such a way that the total soft emission probability, integrated without any constraints, gives one. If the contribution of soft gluons to $\Sigma(v)$ can be written in the closed form in equation (3.60), it is possible to compute it with a Monte Carlo integration procedure, even if the observable cannot be handled analytically [111]. In fact, one just needs to know the function $V(\{\tilde{p}\}, k_1, \dots, k_n)$ for an arbitrary set of soft and collinear emissions. We can elaborate equation (3.60) a bit further to make contact with equation (3.43). The general idea is to split the integration region in the exponent according to whether $V(\{\tilde{p}\}, k)$ is larger or smaller than the observable's value v. This gives

$$\Sigma(v) \simeq e^{-R(v)} \left[e^{-\int^{v} [dk]M^2(k)} \sum_{n=0}^{\infty} \frac{1}{n!} \int \prod_{i=1}^{n} [dk_i]M^2(k_i)\, \Theta\left(v - V(\{\tilde{p}\}, k_1, \dots, k_n)\right) \right], \quad (3.61)$$

where the 'radiator' $R(v)$ is defined as

$$R(v) \equiv \int [dk]M^2(k)\, \Theta\left(V(\{\tilde{p}\}, k) - v\right). \quad (3.62)$$

The radiator exponent corresponds to the Sudakov form factor in equation (3.43). This is a LL function, exponentiates, and factorises from the rest of the distribution. For many observables, the factor that multiplies the exponent of the radiator is a NLL logarithmic function. This is seen most easily in the example of an additive observables, i.e. those satisfying, in the soft-collinear limit

$$V(\{\tilde{p}\}, k_1, \dots, k_n) = \sum_{i=1}^{n} V(\{\tilde{p}\}, k_i). \quad (3.63)$$

The main strategy to isolate NLL contributions is to parametrise each emissions k in terms of $\zeta \equiv V(\{\tilde{p}\}, k)/v$. For additive observables

$$V(\{\tilde{p}\}, k_1, \dots, k_n) = v(\zeta_1 + \zeta_2 + \cdots \zeta_n). \quad (3.64)$$

Since additive observables, for a fixed observable's value v, depend only on the ζ_i, we can freely integrate over all other variables. This implies that we can simplify the integration over the phase space of each emission as follows

$$[dk]M^2(k) = \frac{d\zeta}{\zeta}R'(\zeta v), \qquad R'(v) \equiv -v\frac{d}{dv}R(v). \tag{3.65}$$

By construction, the function $R'(v)$ is a positive-definite monotonic function of v, increasing as v decreases. Therefore, we can rewrite the factor that multiplies the exponent of the radiator in equation (3.61) as follows

$$\begin{aligned} & e^{-\int^{v}[dk]M^2(k)}\sum_{n=0}^{\infty}\frac{1}{n!}\int\prod_{i=1}^{n}[dk_i]M^2(k_i)\,\Theta\left(v - V(\{\tilde{p}\}, k_1, \dots, k_n)\right) \\ &= e^{-\int_{\delta}^{1}\frac{d\zeta}{\zeta}R'(\zeta v)}\sum_{n=0}^{\infty}\frac{1}{n!}\int_{\delta}^{\infty}\prod_{i=1}^{n}\frac{d\zeta_i}{\zeta_i}R'(\zeta_i v)\,\Theta\left(1 - \sum_i \zeta_i\right). \end{aligned} \tag{3.66}$$

In the last line of the above equation, we have introduced the tiny cutoff δ, as we can see that emissions with $\zeta_i < \delta$ cancel against virtual corrections (the negative exponent), up to corrections that vanish as a power of δ. This makes it possible to approximate each $R'(\zeta v)$ with $R'(v) \equiv R'$ in equation (3.66), with corrections of relative order α_s, hence subleading. This gives

$$\begin{aligned} & e^{-\int_{\delta}^{1}\frac{d\zeta}{\zeta}R'(\zeta v)}\sum_{n=0}^{\infty}\frac{1}{n!}\int_{\delta}^{\infty}\prod_{i=1}^{n}\frac{d\zeta_i}{\zeta_i}R'(\zeta_i v)\,\Theta\left(1 - \sum_i \zeta_i\right) \\ &\simeq \delta^{R'}\sum_{n=0}^{\infty}\frac{(R')^n}{n!}\int_{\delta}^{\infty}\prod_{i=1}^{n}\frac{d\zeta_i}{\zeta_i}R'(v)\,\Theta\left(1 - \sum_i \zeta_i\right) \equiv \mathcal{F}(R'). \end{aligned} \tag{3.67}$$

The corrections to the double logarithmic Sudakov exponent $\exp[-R(v)]$ amount to a function of R' only, hence NLL. This implies that, for additive observables, the NLL resummation of soft *and* collinear logarithms amounts to the concise expression

$$\Sigma(v) = e^{-R(v)}\mathcal{F}(R'). \tag{3.68}$$

For additive observables, $\mathcal{F}(R')$ can be computed analytically, and reads

$$\mathcal{F}(R') = \frac{e^{-\gamma_E R'}}{\Gamma(1 + R')}. \tag{3.69}$$

This is of course identical to what one obtains by solving the NLL evolution equation (3.53) for the jet mass, and applying the result to the thrust distribution as in equation (3.56).

Given $V(\{\tilde{p}\}, k_1, \dots, k_n)$, it is possible to establish if equation (3.60) is enough to obtain a NLL resummation for $\Sigma(v)$. Essentially, the observable needs to have the same properties as additive observables, namely one needs to be able to treat multiple soft and collinear emissions as independent, and neglect all emissions that have $V(\{\tilde{p}\}, k) < \delta v$. The set of requirements that $V(\{\tilde{p}\}, k_1, \dots, k_n)$ needs to satisfy are known as 'continuous globalness' (this leads to the absence of non-global logarithms) and 'recursive infrared and collinear (rIRC) safety' [112]. These are more restrictive constraints with respect to standard IRC safety, and are related to the scaling behaviour of $V(\{\tilde{p}\}, k_1, \dots, k_n)$ in the presence of multiple soft and/or collinear emissions. It is possible to show that the NLL resummation for any continuously global rIRC final-state observable is given by a general master formula, obtained by augmenting equation (3.60) with virtual corrections of hard collinear and soft wide-angle origin. These contributions factorise, exponentiate, and can be computed analytically for all observables. The most important feature of rIRC safe observables is that the ingredients of the NLL resummation formula can be extracted by evaluating $V(\{\tilde{p}\}, k_1, \dots, k_n)$ with suitable ensembles of soft and/or collinear emissions. This evaluation can be efficiently performed by a computer, and is encoded in the numerical code Computer Automated Expert Semi-Analytical Resummer (CAESAR) [112]. The CAESAR approach is very effective in handling contributions arising from multiple soft-collinear emissions. These are computed through a simplified parton-shower event generator, that produces an arbitrary number of soft and collinear emissions, and feeds them into a computer subroutine that computes $V(\{\tilde{p}\}, k_1, \dots, k_n)$. However, the cancellation of infrared and collinear singularities between real and virtual corrections has to be performed analytically. The outcome is an exponent that contains virtual corrections, and unresolved real emissions, which give no contribution to $V(\{\tilde{p}\}, k_1, \dots, k_n)$ (for instance, because they are too soft), and is in fact the Sudakov form factor for the observable V. CAESAR is also able to determine if an observable is rIRC safe given its expression in terms of final-state momenta.

It is possible to extend the CAESAR philosophy to NNLL accuracy. This is at the core of the ARES approach [113–115]. The main idea is again to identify unresolved emissions, which do not contribute to the observable in question, and cancel soft and collinear singularities of virtual corrections. The combination of unresolved emissions and virtual corrections can be computed analytically, and the remaining corrections due to resolved real emissions can be generally computed with suitably designed Monte Carlo event generators. The ARES approach has been developed for two-jet [113, 114] and three-jet [116] observables in e^+e^- annihilation. The only missing piece for its extension to an arbitrary process is the general calculation of the contribution of unresolved emissions. In fact, virtual corrections at the needed accuracy are already known for an arbitrary number of emitting legs [117]. While in the original ARES formulation, the combination of unresolved real emissions and virtual corrections is computed in QCD, nothing prevents its calculation in a different theoretical framework, for instance in SCET (see e.g. [118]).

An important message of this section is that having a resummation formula in terms of actual emissions rather than as the solution of a differential equation has the clear advantage that one has a neat picture of what approximations on the multi-parton emission probabilities and phase space are needed to achieve a given logarithmic accuracy. This plays a very important role in establishing the accuracy of parton-shower event generators, the topic of the next section, whose aim is to realistically simulate multiple soft and/or collinear emissions. In fact, one needs to validate the approximations according to which partons are produced by the parton-shower algorithm. This is one of the main benefits of the cross-talk between analytic resummation and the development of parton-shower event generators.

3.2.3 Parton-shower event generators

Let us consider again equations (3.40) and (3.41) for the jet mass distributions J_q and J_g respectively. These equations could be solved iteratively by expanding in the number of real emissions. Keeping terms with up to one emission, for J_q we obtain

$$J_q(Q^2, k^2) = \Delta_q(Q^2, 0)\delta(k^2) + \int_0^{Q^2} \frac{dq^2}{q^2}\Delta_q(Q^2, q^2)\Delta_q(q^2, 0)\Delta_g(q^2, 0) \\ \times \int_0^{1-\epsilon(q^2)} dz\, \frac{\alpha_s}{2\pi} P_{q\to qg}(z)\delta(k^2 - q^2) + \text{terms with more emissions} \tag{3.70}$$

where again $\epsilon(q^2)$ is a cutoff needed to regularise the z integration. The interpretation of this equation is as follows. If there are no splittings, which happens with probability $\Delta_q(Q^2, 0)$, the invariant mass of the jet is zero. If there is exactly one splitting, the jet invariant mass is determined by the total invariant mass of the products of the splitting, and so on. The terms of such iterative expansion correspond to mutually exclusive events with a finite (although not fixed *a priori*) number of emissions. Once we have produced any such event, we can bin the total invariant mass, and the resulting histogram, for a very large number of events, will reproduce $J_q(Q^2, k^2)$. Instead of binning the invariant mass distribution, one could bin another variable, since the variables q^2, z, and possibly a uniformly distributed azimuth ϕ, can be used to reconstruct the kinematics of each event. In equation (3.70) it is not specified how to regularise the Sudakov form factor $\Delta_q(Q^2, 0)$, as well as the q^2 integral. A viable way that is amenable to integration in four dimensions is to declare all emissions with q^2 less than a small cutoff Q_0^2 as unresolved, and hence not contributing to any observable. This amounts in modifying equation (3.70) as follows

$$J_q(Q^2, k^2) = \Delta_q(Q^2, Q_0^2)\delta(k^2) + \int_{Q_0^2}^{Q^2} \frac{dq^2}{q^2}\Delta_q(Q^2, q^2)\Delta_q(q^2, Q_0^2)\Delta_g(q^2, Q_0^2) \\ \times \int_0^{1-\epsilon(q^2)} dz\, \frac{\alpha_s}{2\pi} P_{q\to qg}(z)\delta(k^2 - q^2) + \text{terms with more emissions.} \tag{3.71}$$

The above equation can be simulated with a computer. First, we identify the probability of a quark splitting into a quark and a gluon:

$$dP_{q\to qg} \sim \frac{dq^2}{q^2}\Delta_q(q'^2, q^2)\Theta(q'^2 - q^2)\, dz P_{q\to qg}(z)\frac{\alpha_s}{2\pi}\Theta(1 - \epsilon(q^2) - z). \tag{3.72}$$

Then we generate a value of q^2 according to equation (3.72). If this value is less than Q_0^2, the procedure stops, and no emissions are produced. If $q^2 > Q_0^2$, we have one splitting, and the quark gives rise to another quark and a gluon. The quark branch can split further according to equation (3.72). The gluon branch can split analogously either into two gluons with probability

$$dP_{g\to gg} \sim \frac{dq^2}{q^2}\Delta_g(q'^2, q^2)\Theta(q'^2 - q^2)\, dz P_{g\to gg}(z)\frac{\alpha_s}{2\pi}\,\Theta(1 - \epsilon(q^2) - z)\,\Theta(z - \epsilon(q^2)). \quad (3.73)$$

or to a quark–antiquark pair (both massless, with n_f flavours) with probability

$$dP_{g\to q\bar{q}} \sim \frac{dq^2}{q^2}\Delta_g(q'^2, q^2)\,\Theta(q'^2 - q^2)\, dz\, n_f\, P_{g\to q\bar{q}}(z)\frac{\alpha_s}{2\pi}. \quad (3.74)$$

For simplicity here we have assumed that all n_f massless flavours contribute in the same way to the observable, but one can keep track of the contribution of each individual flavour. For the first splitting, the scale q'^2 is equal to the hard scale Q^2, and for subsequent splittings it is the value of q^2 of the previous splitting, see equation (3.71). At each splitting, if q^2 is less than the cutoff Q_0^2, the procedure stops, and any physical observable can be computed on the emissions present at that stage. One can see that each emission corresponds to the jet branching further. In practice, instead of taking into account all branches simultaneously, one follows a given branch until $q^2 < Q_0^2$, then another, until all branches have produced values of q^2 less than Q_0^2.

The procedure we have outlined here is a simplified example of a parton-shower event generator. These computer programs simulate successive collinear splittings, ordered in some variable q^2 proportional to the splitting opening angle θ^2 (q^2 can be the invariant mass of the parent parton, but it does not need to be so). The ith splitting in the cascade, of type $a \to bc$, is generated with probability

$$dP_{a\to bc} \sim \frac{dq^2}{q^2}\Delta_a(q'^2, q^2)\,\Theta(q'^2 - q^2)\; dz P_{a\to bc}(z)\frac{\alpha_s}{2\pi}\,\Theta(1 - \epsilon(q^2) - z)\,\Theta(z - \epsilon(q^2)), \quad (3.75)$$

with $\Delta_a(Q^2, Q_0^2)$ the generalised Sudakov form factor

$$\Delta_a(Q^2, Q_0^2) = \exp\left[-\int_{Q_0^2}^{Q^2}\frac{dq^2}{q^2}\int_{\epsilon(q^2)}^{1-\epsilon(q^2)} dz\,\frac{\alpha_s}{2\pi}\sum_b P_{a\to bc}(z)\right]. \quad (3.76)$$

In the above equation, the flavour of parton c is determined by a and b, hence that index is not summed over. The different ordering variable is reflected in the boundary of the z integration, which have been left intentionally unspecified in equation (3.75). Once q^2 and z have been generated, one can reconstruct the full kinematics of the splitting, i.e. the momenta p_a, p_b and p_c. During this procedure the ordering variable decreases, until it reaches a minimum value, where the procedure stops, leaving a set of final-state partons. These partons are then transformed into final-state hadrons by means of a hadronisation model, and these hadrons are used as inputs for the physical observable under consideration. For instance, if we wish to compute the sum of the invariant masses of the two hemispheres in e^+e^- annihilation, we first produce a quark and an anti-quark, let them split subsequently and produce a set of final state partons, and then hadrons. Then, the hadron momenta are separated in two hemispheres, and the sum of the invariant masses of the two hemispheres is calculated. A parton-shower program returns also the weight of each event, which can then be suitably used to produce histograms. Parton-shower event generators are among the most widely used tools in high-energy physics. Their strength is twofold:

- they describe correctly at tree level an arbitrary number of collinear splittings and
- they produce realistic simulated events, whose final states correspond to actually observed particles.

The last point is particularly important for experimental analyses. The final-state momenta produced by a parton-shower event generators can be sent directly to a detector simulator, that returns the signals that would be observed in an actual detector, given the input momenta. It is these reconstructed momenta that constitute the input for physics analyses in high-energy physics experiments. Therefore, the ideal parton-shower event generator should be able to simulate events with the same probability as they would occur in reality. Of course, this is not possible from a theoretical point of view, hence some approximations need to be employed. However, current event generators do give a satisfactory description of experimental data. This success relies on several improvements with respect to the naïve formulation we have presented above.

Branching algorithms. Many improvements concern the modification of the splitting probability of equation (3.75), so as to account for the largest number of logarithmically enhanced contributions. As we have already remarked a number of times, collinear logarithms can be accurately taken into account by evolution equations such as equations (3.40) and (3.41). The problem remains soft emissions which, as we have discussed already in section 3.2.2, do not fit a simple probabilistic picture.

There are essentially two different possibilities to solve this problem. The first is to improve the treatment of collinear splitting to take into account coherence of soft radiation, i.e. the fact that a soft gluon at large angle sees only the total colour charge of all partons at smaller angles. This is the strategy employed by the original version of the widely used event generator PYTHIA [119], which uses the invariant mass as ordering variable for collinear splittings, and corrects the phase space of emitted partons to account for coherence. Another parton-shower algorithm uses, implemented in most recent versions of PYTHIA [120], uses the relative transverse momentum of each splitting as an ordering variable [121]. A competing algorithm the evolution equation (3.47) as a starting point, i.e. employs the splitting angle as an evolution variable. This is the basis of the so-called 'coherent branching' algorithm, implemented in the event generator HERWIG [122–124]. Other branching algorithms, like the ones implemented in HERWIG++ [125] and SHERPA [126], are similar in philosophy but differ in the coverage of the phase space of the emitted partons.

Angular ordering implies that soft and collinear emissions widely separated in angle are emitted independently from hard partons. For processes with two emitting hard partons that form a colour singlet, i.e. with zero total colour charge, for instance a quark–antiquark pair produced via e^+e^- annihilation, the algebra of colour matrices is so simple that independent emission accounts for all soft contributions, including those at large angles. Therefore, as explained at the end of section 3.2.2, an angular-ordered parton-shower will provide NLL accurate predictions for any continuously global rIRC safe two-jet observable in e^+e^- annihilation. However, it has been shown that angular ordering does not achieve NLL accuracy for non-global observables [127]. If one wishes to describe those observables as well, at least in the large-N_c limit, one needs to implement a parton branching that achieves the iterative solution of the non-linear equation that resums non-global logarithms [88, 91, 92]. The resulting branching algorithm involves 'dipoles' as basic objects. In fact, a soft gluon is emitted by pairs of partons, each pair being a dipole. When a gluon is emitted, a new dipole is formed, which can in turn produce one more dipole, and so on. Variants of such algorithms, called 'dipole' showers, differ in the way in which they accommodate collinear splittings, and on how they reconstruct the event kinematics. The first program implementing these ideas was ARIADNE [128], a widely used tool in e^+e^- annihilation. Dipole showers are the basis of the Monte Carlo event generator VINCIA [129], and the DIRE dipole shower [130] is implemented in the current version of PYTHIA [120]. It was generally believed that, since gluons are emitted with the correct probability, coherence was automatically taken into account by dipole shower algorithms. However, the way an algorithm behaves in the collinear region affects the probability of emissions of multiple soft gluons widely separated in angle, inducing non-trivial correlations that are not present in QCD squared amplitudes [131]. This observation is the basis of the project PanScales [132], that constructed a dipole shower that achieves NLL accuracy for both global and non-global observables. This is obtained through a careful selection of the evolution variable, and on the strategy to reconstruct the event kinematics from the dipole variables. The original formulation of the PanScales algorithm, designed for e^+e^- annihilation, has been also extended to hadron collisions [133].

We remark however that dipole shower in general are accurate only in the large-N_c limit. Moving towards finite-N_c involves typically including some contributions that are suppressed by inverse powers of N_c. We note however some more ambitious programmes of including colour effects to all orders starting from QCD amplitudes, instead of branching probabilities (see e.g. [134, 135]).

Matching of fixed-order and parton showers. Another relevant direction of improvement for parton-shower event generators is in the generation of the hard event. The branching algorithms described so far start from a given set of hard partons that initiate collinear branchings. In their basic version, all event generators produce these partons at leading order in QCD perturbation theory, and all subsequent partons are produced through the branching algorithm. But this does not properly account for events in which new hard partons are produced. One way to deal with this problem is merging procedures, the pioneering ones named CKKW [136] and MLM [137], in which a jet-resolution parameter, usually referred to as the 'merging scale', is introduced. When a new parton is emitted, if its distance (suitably defined) with respect to the other partons is below the resolution parameter, it will be considered as a soft-collinear parton, otherwise it will be considered as a new hard parton, and the event will be given the correct lowest-order probability for the production of a new hard parton. With this procedure one can merge different jet multiplicities. This is needed in all analyses in which the signal we are interested in produces very energetic jets. For instance, supersymmetric particles, in particular squarks and gluinos, are expected to produce many jets at high transverse momentum through their decays into coloured particles. Therefore a correct estimate of multi-jet emission probability is essential for the estimation of the QCD background in searches for such particles. There are cases in which the particle we are looking for is typically accompanied by at most one extra jet. In this case, it is important to precisely determine the rate of production of the particle, i.e. compute it at least at NLO. There are procedures through which it is possible to generate NLO events with a modified weight, so that they can be used as the starting point of a parton-shower event generator. The key point here is to avoid double counting, as an extra emission can be produced by the parton-shower event generator. Another wanted feature is that the inclusive sum of all the weights gives the correct cross section at NLO accuracy. These problems have been solved through well-established 'NLO matching' procedures. The most popular ones are those implemented in the public programs MC@NLO [138] and POWHEG [139]. Both generate hard events that can be suitably interfaced with any parton-shower event generator. More sophisticated procedures, named MEPS@NLO [140] and FxFx [141] successfully combine NLO matching with merging of higher jet multiplicities.

With the knowledge we have acquired so far we are able to assess the performance of different parton-shower event generators and fixed-order calculation in describing selected set of data. For instance, in figure 3.10 we consider the transverse momentum distribution of the leading jet (the one with the largest transverse momentum) produced in association with a Z boson, measured by the ATLAS collaboration a the LHC with $\sqrt{s} = 7$ TeV [142] (left) and $\sqrt{s} = 13$ TeV [5] (right). It is instructive to observe how the tools employed to describe the data have changed with time, reflecting our improved understanding of the dynamics underlying jet production.

Let us consider the data at $\sqrt{s} = 7$ TeV. These are compared to the NLO program BLACKHAT+SHERPA, and the parton-shower event generators MC@NLO, SHERPA and ALPGEN. MC@NLO is set up in such a way that the total Z production cross section is correct at NLO. This means that the first jet produced in the branching is correctly produced at LO accuracy, whereas all other jets are produced by the parton-shower, which is accurate in the collinear limit only. In fact, the calculation of MC@NLO has essentially the same physics content as the LO calculation illustrated in section 3.1.1, which explains why it underestimates the

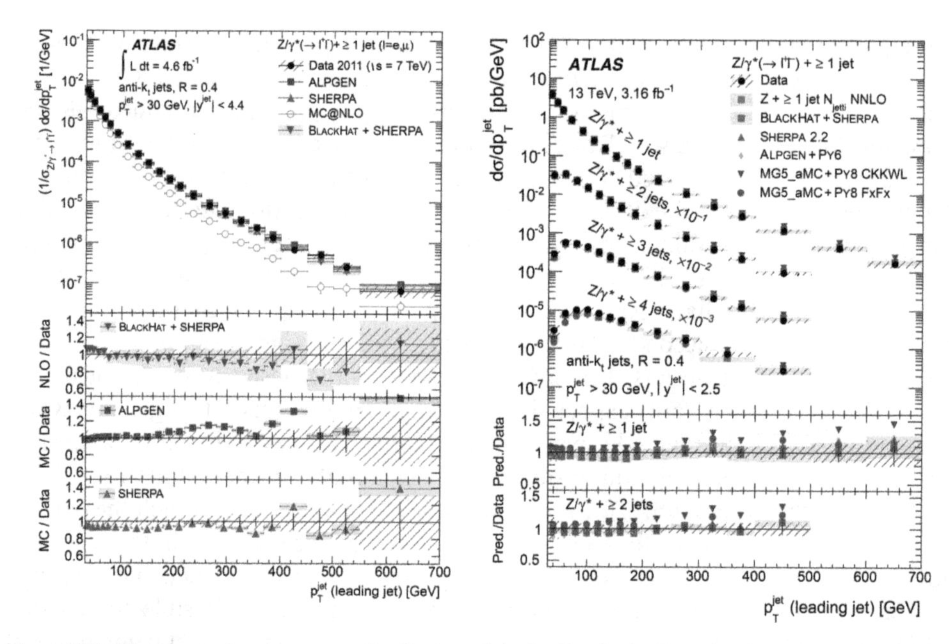

Figure 3.10. The transverse momentum distribution of the leading jet in Z production, as measured by the ATLAS collaboration at the LHC with $\sqrt{s} = 7$ TeV (left, reproduced from [142] (2013), with permission of Springer) and $\sqrt{s} = 13$ TeV (right, © [5] (2017), with permission of Springer).

data, especially at high transverse momentum. The correct normalisation is accounted for, within theoretical uncertainties, by the NLO calculation provided by BLACKHAT+SHERPA. Note that SHERPA and ALPGEN, which implement the leading-order multijet merging procedures of [136] and [137] respectively, give the correct normalisation in spite of the fact that they do not have the full virtual corrections, but only the approximate value given by the Sudakov form factor. The message we extract is that, for this process, the main role of virtual corrections is that of cancelling divergences of real-emission contributions, and that leftover finite terms are quite small, and contained within theoretical uncertainties.

Considering the data on the right-hand panel of figure 3.10, corresponding to $\sqrt{s} = 13$ TeV, we observe that they are not compared to plain MC@NLO any more, but with two versions of that program that include multi-jet merging. The first, labelled MG5_aMC CKKWL, corresponds to MADGRAPH_aMC@NLO [13] interfaced to PYTHIA version 8.1 [143], and CKKWL multi-jet merging [144], a version of the pioneering CKKW merging procedure [136] tailored to dipole showers. The other, labelled MG5_aMC FxFx, is MADGRAPH_aMC@NLO interfaced to PYTHIA with FxFx [141] multi-jet merging, which provides NLO matching up to two jets. Other parton-shower event generators used for the comparison are SHERPA, version 2.2 [145] and ALPGEN [146] interfaced to PYTHIA version 6.4 [119]. Both implement multi-jet merging, namely SHERPA the CKKW procedure [136] and ALPGEN the MLM procedure [137]. We observe that multi-jet merging is crucial to obtain a satisfactory description of data. We note that

MG5_aMC CKKWL tends to overshoot the data at large transverse momentum, and this unwanted feature is removed when upgrading to FxFx [141] merging. In the same plot, data is compared to fixed order predictions, both at NLO obtained with BLACKHAT+SHERPA, and at NNLO obtained with a phase-space slicing subtraction scheme based on the variable N-jettiness [147]. We observe that both NLO and NNLO calculations describe data quite well. The NNLO prediction has smaller theoretical uncertainties, due to the fact that the relevant two-jet multiplicity is described at NLO instead of LO. The plot also shows data corresponding to higher jet multiplicities, compared to parton-shower event generators. They all describe data quite well, except MG5_aMC CKKWL, which tends to overshoot the data at large transverse momentum.

At higher luminosities, higher regions of transverse momentum become available (see e.g. [148] for a recent experimental analysis). There one probes energy scales that are much larger than the mass of the Z boson. This implies that electro-weak corrections, such as those induced by the radiation of soft Z bosons, cannot be neglected any more, and the simplified discussion we have presented so far needs to be improved.

The leading jet in Z events at high transverse momentum is described satisfactorily by both fixed-order calculations and parton-shower event generators with multi-jet merging. This is because this observable is sensitive only to the presence of hard partons, with soft and collinear emissions essentially cancelling at all orders against virtual corrections. Parton-shower event generators are instead crucial for describing observables that are sensitive to multiple soft and collinear emissions, as is the case of the distribution in the transverse momentum p_T^Z of a Z boson at low p_T^Z. A comparison of fixed-order calculations [149, 150] with experimental data (see figure 3.11, left-hand panel) show that theory predictions break down at small values of the Z transverse momentum, due to the presence of large logarithms $\ln(M_Z/p_T^Z)$. This breakdown is not particularly severe as the predictions correspond to Z production plus one jet at the high NNLO accuracy. Parton-shower event generators (see figure 3.11, right-hand panel) describe data much better, although their

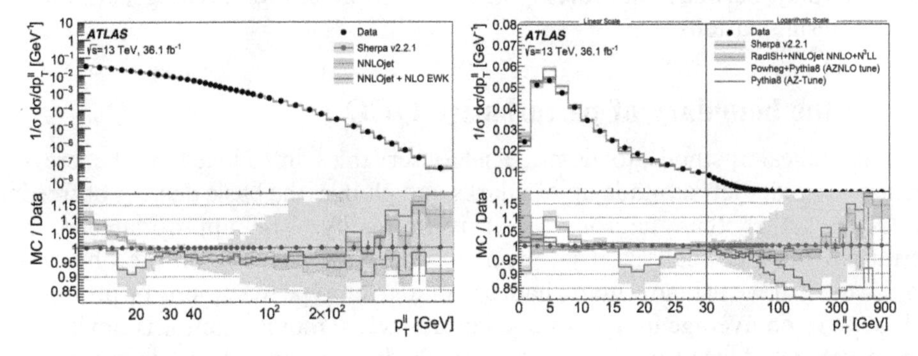

Figure 3.11. Comparison of ATLAS data for the transverse momentum distribution of a Z boson to fixed-order calculations (left) and parton-shower event generators and resummed predictions (right).

normalisation does not perfectly agree with data at small values of p_T^Z. In the same panel, resummed predictions are shown, obtained at very high accuracy (N^3LL) with the program RadISH [152, 153]. Observables like the Z transverse momentum distribution have then significant scope in assessing the performance of various approaches to parton shower, as well as for tuning the parameters associated with the shower algorithms.

In addition to improvements in branching algorithms, and the matching of more and more processes at NLO, more sophisticated parton-shower event generators aim at the merging of different jet multiplicities, in such a way that the rate of each multiplicity is NLO accurate [154–156]. For instance, through the MiNLO procedure, it became possible to merge for the first time Higgs events with zero and one jet at NLO, without any intermediate merging scale as in traditional multi-jet merging. Furthermore, with a simple re-weighting of zero-jet events, events generated with MiNLO add up to give the total Higgs cross section at NNLO [157]. MiNLO with this rescaling gets renamed as MiNNLO. The MiNNLO procedure represents therefore the first example of a parton shower matched to NNLO. Matching parton-showers with NNLO represents the frontier of matching of parton-shower with event generators. Notably, the MiNNLO procedure has been further applied to the cases of Z, di-boson and $t\bar{t}$ production (see e.g. [158]).

To conclude, parton-shower event generators are leaving their traditional role of tools for approximate simulation of collider events, to become more and more precision tools, able to predict also jet distributions with a reliable normalisation, and to account to some extent for their own theoretical uncertainties. This has been possible also because parton-shower generators have incorporated many theoretical advances in analytic calculations. For instance, to achieve NLO accuracy for two different jet multiplicities in Higgs production it was necessary to improve the Sudakov form factor in such a way that logarithmically enhanced contributions in the transverse momentum of the Higgs were described at very high accuracy. Fortunately, these logarithms were known already from analytical calculations. Similarly, the NLL accuracy of the PanScales generator benefited crucially from the existence of a general framework for NLL resummations. This shows once more the vital interplay between the development of parton-shower event generators and analytic resummations.

3.3 At the boundary of perturbative QCD

The theoretical methods to describe jet observables introduced in the previous sections consider jets as made up of quarks and gluons, the basic degrees of freedom of QCD, rather than hadrons, as it is in reality. The appropriateness of a perturbative QCD analysis of jet properties depends crucially on the observables we consider. If we ask how many hadrons of a given type (e.g. pions or protons) we will observe on average in a jet of a given transverse momentum, it is obvious that the answer will depend on the details of the hadronisation process. However, if we look at the distribution in the transverse momentum of a jet originating from an energetic quark, we expect that the reshuffling of final-state momenta due to hadronisation will not change dramatically the total transverse momentum of the jet. Let us make this argument more rigorous. Suppose we have a generic cross section σ that depends on a typical scale Q, (e.g. the transverse momentum of a jet), and on the masses of the quarks involved m_q. Collinear divergences are regulated by the quark masses, and we introduce also a fake gluon mass m_g to regulate soft divergences. For $Q^2 \gg m_q^2, m_g^2$, the dependence of σ on the relevant scales involved is given by

$$\sigma(\alpha_s(\mu), \mu^2, Q^2, \{m_i^2\}) = \frac{1}{Q^2} F\left(\frac{Q^2}{\mu^2}, \left\{\frac{m_i^2}{\mu^2}\right\}, \alpha_s(\mu)\right), \quad i = q, g, \tag{3.77}$$

with μ the renormalisation scale arising after the renormalisation of ultraviolet divergences. If soft and collinear divergences cancel completely between real and virtual corrections, then this cross section can be computed in the limit in which all masses are set to zero. More precisely, the following limit exists

$$\lim_{Q\to\infty} Q^2 \sigma(\alpha_s(\mu), \mu^2, Q^2, \{m_i^2\}) = \hat{F}\left(\alpha_s(\mu), \frac{Q^2}{\mu^2}\right). \tag{3.78}$$

For finite Q, we can write [159]

$$F\left(\alpha_s(\mu), \frac{Q^2}{\mu^2}, \left\{\frac{m_i^2}{\mu^2}\right\}\right) = \hat{F}\left(\alpha_s(\mu), \frac{Q^2}{\mu^2}\right)\left[1 + \mathcal{O}\left(\frac{m_i^2}{Q^2}\right)^p\right], \tag{3.79}$$

where p is a positive power. In practice, this means that, when $Q^2 \gg m_q^2, m_g^2$, a good estimate for F is provided by its asymptotic value $\hat{F}$, since we are losing only power-suppressed corrections. It is these corrections that are responsible for hadronisation effects. Therefore, we can say that, when the typical hard scale of a process is much bigger than the small quark and gluon masses, hadronic cross sections can simply be computed using the quark–gluon language of perturbative QCD. The implication in terms of jet physics is that, when Q is extremely large, radiation from hard partons is forced to be extremely collimated, hence we can safely assume that one jet is obtained per hard parton produced.

For moderate energies, however, hadronisation corrections do play a role and need to be quantified. This is normally achieved through hadronisation models available in parton-shower event generators. These models are phenomenological tools, encoding heuristic, though not simplistic, ideas of how hadronisation works. They contain many parameters, which have to be determined by comparing predictions for suitable observables, sensitive to hadronisation (like hadron multiplicities inside jets, event-shape mean values and distributions), to experimental data. Within this framework, hadronisation corrections to IRC safe observables (e.g. jet cross sections) can be evaluated by simply taking the ratio of the predictions of a parton-shower event generator at hadron level and at parton level (with hadronisation switched off). Such a procedure is not inappropriate, in that the fact that hadronisation corrections are power-suppressed suggests that the phenomena underlying the formation of hadrons are largely decoupled from the energetic parton branching giving rise to the energy-momentum pattern of jets. The use of such an approach to precision measurements, e.g. the determination of the strong coupling, is appropriate as long as perturbative uncertainties are larger than those arising from different hadronisation models. With the perturbative accuracy available nowadays, the validity of this method has been questioned [160]. The attention has then turned more and more towards analytic models of hadronisation corrections, which we will briefly present in the following.

Local parton–hadron duality. As stated before, it is possible to estimate the size of hadronisation corrections to a certain extent with analytic methods. Fully inclusive quantities, like the total cross section $e^+e^- \to$ hadrons, can be expressed in terms of Fourier transforms of expectation values of products of operators in quantum field theory, for instance

$$\sigma(e^+e^- \to \text{hadrons}) \sim \int d^4x \, e^{iqx} \langle 0 | J(x) \cdot J(0) | 0 \rangle, \tag{3.80}$$

where the four-vector $J^\mu(x)$ is a local operator representing the conserved quark electromagnetic current, q is the total four-momentum of the incoming e^+e^- pair, and $|0\rangle$ is the vacuum of the theory. The lack of knowledge of the vacuum is the major stumbling block in the calculation of the right-hand side of equation (3.80), so some assumptions and approximations have to be employed. First, the quantity in equation (3.80) is a function of the invariant $Q^2 = q^2$, and can be expanded in inverse powers of Q^2. The leading term in $1/Q^2$, the so-called 'leading-twist' term, gives precisely what we would obtain in massless perturbative QCD. In a sense, it corresponds to approximating the vacuum with that of a quasi-free theory, obtained as a small perturbation of the vacuum of a free theory. Higher powers of $1/Q^2$, called 'higher-twist' terms, can be systematically accounted for in this expansion, and consist of products of coefficients, calculable using the quark–gluon language, and expectation values of local operators, which have to be extracted from data, or computed with non-perturbative methods, e.g. discretising QCD on a lattice. Notice that this expansion, called operator product expansion (OPE), is rigorously defined for $Q^2 < 0$, and must necessarily be truncated at a given order. Its analytic continuation into the physical region $Q^2 > 0$ misses completely non-perturbative effects such as the production of hadronic resonances, but captures very well the continuum background in between such resonances (see e.g. [161] for the discussion of this effect through an explicit toy model). This analytic continuation corresponds physically to the assumption that at high energies hadronic physics is mostly accounted for by the dynamics of quarks and gluons, with hadronisation giving just a reshuffling of parton momenta into the observed hadrons. This hypothesis is known as local parton–hadron duality (LPHD), and is assumed to hold also for non-inclusive observables for which an OPE does not exist (see e.g. [161, 162]). LPHD is the philosophy underlying parton-shower event generators, which produce quarks and gluon until a cut-off value in the evolution variable is reached, after which a hadronisation model transforms all produced partons into hadrons [122, 163].

Hadronisation corrections through an effective coupling. An interesting development of LPHD is the idea of describing hadronisation through the notion of a QCD effective coupling at low momentum scales [164]. There are observables like event-shape distributions in which leading hadronisation corrections are due to soft hadrons, at large angles with respect to the leading jets. Despite the fact that, at a fixed order in perturbation theory, gluons emitted in this kinematic region give a very small contribution, it turns out that the perturbative series as a whole is factorially divergent. To overcome this problem, the introduction of a merging scale μ_I has been suggested, such that only emissions with energy above μ_I are treated perturbatively, and in fact they lead to a convergent perturbative expansion [165, 166]. Emissions with transverse momenta below μ_I are treated as soft hadrons, having the same emission probability as soft gluons, except for the coupling which is a non-perturbative extension of the CMW coupling introduced in equation (3.47). For instance, consider a generic two-jet event shape $V(\{\tilde{p}\}, k_1, \dots, k_n)$ in e^+e^- annihilation, to which a non-perturbative ultra-soft emission at large angles contributes by an amount

$$\delta V(\{\tilde{p}\}, k) \simeq (1-z) f_V(\theta), \tag{3.81}$$

with $(1-z)Q/2$ the energy of the non-perturbative gluon, and θ its angle with the respect to the emitting quark or antiquark. In the two-jet region, where only soft emissions are present, the contribution of this emission to the probability $\Sigma(v)$ that $V(\{\tilde{p}\}, \{k_i\}) < v$ is approximately given by

$$\begin{aligned}\delta\Sigma_{\mathrm{NP}}(v) \simeq 2C_F \int_0^1 \frac{dz}{1-z} \frac{\alpha_s^{\mathrm{NP}}[(1-z)\theta Q]}{\pi} \int \frac{d\theta^2}{\theta^2} \Theta(\mu_I - (1-z)\theta Q) \\ \times e^{-\int [dk] M^2(k)} \sum_{n=0}^{\infty} \frac{1}{n!} \int \prod_{i=1}^{n} [dk_i] M^2(k_i) \\ \times \{\Theta[v - V(\{\tilde{p}\}, k_1, \dots, k_n) - \delta V(\{\tilde{p}\}, k)] \\ - \Theta[v - V(\{\tilde{p}\}, k_1, \dots, k_n)]\},\end{aligned} \tag{3.82}$$

where the difference in step functions represents the fact that the ultra-soft emission k can be either real or virtual. Notice that all emissions except k have transverse momenta much bigger than μ_I, and can be then considered as quarks and gluons. Since δV is small with respect to v, one can expand the theta function and obtain

$$\begin{aligned}\delta\Sigma_{\mathrm{NP}}(v) \simeq -2C_F \int_0^1 \frac{dz}{1-z} \frac{\alpha_s^{\mathrm{NP}}[(1-z)\theta Q]}{\pi} \int \frac{d\theta^2}{\theta^2} \delta V(\{\tilde{p}\}, k) \\ \Theta\left(\mu_I - (1-z)\theta Q\right) \\ \times e^{-\int [dk] M^2(k)} \sum_{n=0}^{\infty} \frac{1}{n!} \int \prod_{i=1}^{n} [dk_i] M^2(k_i)\, \delta(v - V(\{\tilde{p}\}, k_1, \dots, k_n)) \\ \equiv -\langle \delta V \rangle_{\mathrm{NP}} \frac{d\Sigma_{\mathrm{PT}}(v)}{dv},\end{aligned} \tag{3.83}$$

where by $\Sigma_{\rm PT}(v)$ we denote the perturbative event-shape distribution, i.e. that computed using quarks and gluons with transverse momenta well above μ_I. The contribution in equation (3.83) represents the leading hadronisation correction to $\Sigma(v)$, including which we have

$$\Sigma(v) \to \Sigma_{\rm PT}(v) - \langle \delta V \rangle_{\rm NP} \frac{d\Sigma_{\rm PT}(v)}{dv} \simeq \Sigma_{\rm PT}(v - \langle \delta V \rangle_{\rm NP}). \tag{3.84}$$

Since $\langle \delta V \rangle_{\rm NP} \sim \mu_I / Q$, hadronisation corrections to such final-state observables amount in a power-suppressed shift in the corresponding distributions. This is actually observed in data, as illustrated for instance in the left-hand panel of figure 3.12. There one sees that the perturbative resummed distribution for one minus the thrust τ (the green dashed curve) has a shape that is roughly consistent with the data, but it needs to be shifted in order to be able to describe the ALEPH data [167] (see the red curve, containing leading hadronisation corrections). The fact that non-perturbative corrections are expressed in terms of the same integral of the universal coupling $\alpha_{\rm s}^{\rm NP}$ makes it possible to extract this quantity from data. In practice, one introduces the non-perturbative parameter $\alpha_0(\mu_I)$, defined as [169]

$$\alpha_0(\mu_I) \equiv \int_0^{\mu_I} \frac{dk}{\mu_I} \alpha_{\rm s}(k), \tag{3.85}$$

with $\alpha_{\rm s}$ the full coupling, valid both in the perturbative and in the non-perturbative regime. By looking at the exact definition of the full coupling $\alpha_{\rm s}$ (see [169] for details), we can then recast the non-perturbative correction to event shape distributions in equation (3.83) in terms of $\alpha_0(\mu_I)$ using the relation

$$\int_0^{\mu_I} dk\, \alpha_{\rm s}^{\rm NP}(k) = \frac{4\mu_I}{\pi}\left[\alpha_0(\mu_I) - \alpha_{\rm s}(Q) + \mathcal{O}(\alpha_{\rm s}^2)\right]. \tag{3.86}$$

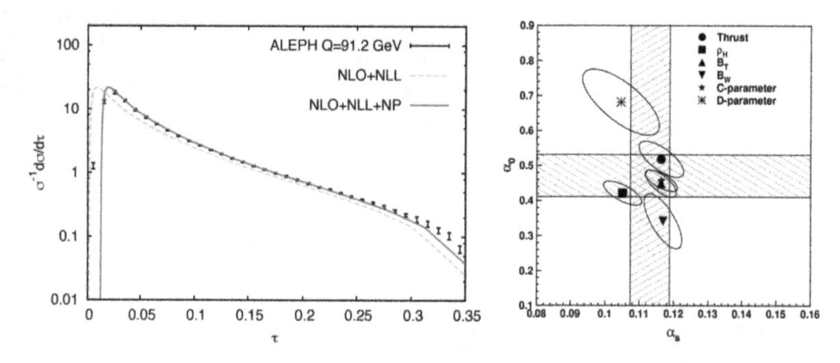

Figure 3.12. Left: theoretical predictions for the distribution in one minus the thrust τ, without [87] (green, dashed) and with (red, solid) non-perturbative (NP) hadronisation corrections [165], compared to ALEPH data [167]. The corresponding theoretical predictions have been matched by the author to exact NLO provided by the program EVENT2 [27]. Right: simultaneous fit of $\alpha_{\rm s}(M_Z)$ and $\alpha_0(2\ {\rm GeV})$ performed by the L3 collaboration [168]. Reprinted from [168], Copyright (2004), with permission from Elsevier.

In practice, one performs simultaneous fits of $\alpha_0(\mu_I)$ and $\alpha_s(M_Z)$. These fits, an example of which can be found in the right-hand plot of figure 3.12, show that α_0 is approximately universal, thus confirming the appropriateness of the effective coupling approach to describe hadronisation corrections to jet observables. It has to be noted that other approaches give the same result: leading power-suppressed corrections to event shapes are of order $1/Q$, and can be expressed in terms of a single universal parameter.

The analytic model described so far is oversimplified. In fact, the non-perturbative correction in equation (3.83) is computed starting from two-jet configurations, which are the dominant ones for small v. However, most fits of the strong coupling are performed in the three-jet region, where v is not small. A recent important development involved the calculation of non-perturbative corrections to event-shape distributions in the three-jet region [170]. The main result is that the same parameter governing non-perturbative corrections in the two-jet region can be used in the three-jet region as well. This opens up new possibilities to improve the description of event-shape distributions, and consequently determinations of the strong coupling. Another improvement would result from the calculation of leading hadronisation corrections to to jet rates, which currently are not known. Phenomenological studies with parton-shower event generators show that these are much smaller than those of event shapes. This is yet another reason why defining observables in terms of jets is preferred to using individual hadrons.

Underlying event. In hadron collisions there is an additional source of non-perturbative corrections, the so-called underlying event (UE). When two hadrons collide at high energies, they break apart, and a single parton is extracted out of each of them. The remnants have themselves a colour charge, and can therefore interact strongly. A rare occurrence is a secondary hard scattering of two partons from the remnants. This is called double-parton scattering, and gives rise to a secondary hard event, for instance the production of two or more jets with high transverse momentum. The most likely situation is that low-energy collisions occur, producing a cloud of low transverse-momentum hadrons, with a distribution that is roughly uniform in rapidity and azimuth. In practice, it is difficult to distinguish between the two situations, because there might be more than one secondary collision, and some secondary collisions will produce such low transverse momentum jets that the resulting event cannot be distinguished from a uniform hadron background (see [171] for a review of the state of the art of multi-parton collisions). This is why many parton-shower event generators, e.g. PYTHIA, model the underlying event as multi-parton interactions (MPI), where proton remnants can undergo one or more secondary collisions [172]. The original version of HERWIG produces soft hadrons uniformly in rapidity and azimuth [123, 124]. However, since this does not account properly for hadron production in the presence of the UE, the package JIMMY, which encodes a model for MPI, has been devised to be interfaced with HERWIG [173]. The main quantity one is interested in when studying jet physics is ρ, the average transverse momentum per unit rapidity and unit azimuth produced by the UE. How this quantity can be extracted from data, and how its fluctuations from one event to the other can be assessed, are still the subject of debate, and beyond the scope of this book (see e.g. [174, 175] for proposals). An analogous quantity ρ_{PU} represents the average transverse momentum per unit rapidity and unit azimuth produced by pile-up (PU) events, secondary hadronic collisions that occur at each beam crossing. UE and PU are unavoidable in hadronic collisions, and contaminate any jet observable. It is therefore important to quantify how much they affect physical observables, and find strategy to assess, and hopefully subtract, their contribution. Some general strategies, such as the method of jet areas, have been discussed in section 2.2. Jets with large radius, such as the ones employed for the search of heavy boosted objects, require event further methods to eliminate a uniform background. This will be one of the main topics of discussion of the next chapter, which deals precisely on how jets can be exploited to look for new phenomena.

References

[1] Ellis R K, Stirling W J and Webber B R 2011 *QCD and Collider Physics* vol 8 (Cambridge: Cambridge University Press), p 2

[2] Dissertori G, Knowles I G and Schmelling M 2003 *Quantum Chromodynamics: High Energy Experiments and Theory* (Oxford: Oxford University Press)

[3] Pfeifenschneider P *et al* (JADE, OPAL Collaboration) 2000 QCD analyses and determinations of alpha(s) in $e^+ e^-$ annihilation at energies between 35-GeV and 189-GeV *Eur. Phys. J.* C **17** 19–51

[4] Campbell J and Neumann T 2019 *J. High Energy Phys.* **12** 034

[5] Aaboud M *et al* (ATLAS Collaboration) 2017 Measurements of the production cross section of a Z boson in association with jets in pp collisions at $\sqrt{s} = 13$ TeV with the ATLAS detector *Eur. Phys. J.* C **77** 361

[6] Butterworth J *et al* 2016 PDF4LHC recommendations for LHC Run II *J. Phys.* G **43** 023001

[7] Nogueira P 1993 Automatic Feynman graph generation *J. Comput. Phys.* **105** 279–89

[8] Hahn T 2001 Generating Feynman diagrams and amplitudes with FeynArts 3 *Comput. Phys. Commun.* **140** 418–31

[9] Belyaev A, Christensen N D and Pukhov A 2013 CalcHEP 3.4 for collider physics within and beyond the standard model *Comput. Phys. Commun.* **184** 1729–69

[10] Press W H, Teukolsky S A, Vetterling W T and Flannery B P *Numerical Recipes The Art of Scientific Computing* (Cambridge: Cambridge University Press)

[11] Lepage G P 1978 A new algorithm for adaptive multidimensional integration *J. Comput. Phys.* **27** 192

[12] Stelzer T and Long W F 1994 Automatic generation of tree level helicity amplitudes *Comput. Phys. Commun.* **81** 357–71

[13] Alwall J, Frederix R, Frixione S, Hirschi V, Maltoni F, Mattelaer O, Shao H S, Stelzer T, Torrielli P and Zaro M 2014 The automated computation of tree-level and next-to-leading order differential cross sections, and their matching to parton shower simulations *J. High Energy Phys.* **07** 079

[14] De Causmaecker P, Gastmans R, Troost W and Wu T T 1982 Multiple bremsstrahlung in gauge theories at high-energies. 1. General formalism for quantum electrodynamics *Nucl. Phys.* B **206** 53–60

[15] Berends F A, Kleiss R, De Causmaecker P, Gastmans R, Troost W and Wu T T 1982 Multiple bremsstrahlung in gauge theories at high-energies. 2. Single bremsstrahlung *Nucl. Phys.* B **206** 61–89

[16] Parke S J and Taylor T R 1986 An amplitude for *n* gluon scattering *Phys. Rev. Lett.* **56** 2459

[17] Witten E 2004 Perturbative gauge theory as a string theory in twistor space *Commun. Math. Phys.* **252** 189–258

[18] Cachazo F, Svrcek P and Witten E 2004 MHV vertices and tree amplitudes in gauge theory *J. High Energy Phys.* **09** 006

[19] Kosower D A 2005 Next-to-maximal helicity violating amplitudes in gauge theory *Phys. Rev.* D **71** 045007

[20] Berends F A and Giele W T 1988 Recursive calculations for processes with *n* gluons *Nucl. Phys.* B **306** 759–808

[21] Caravaglios F and Moretti M 1995 An algorithm to compute Born scattering amplitudes without Feynman graphs *Phys. Lett.* B **358** 332–8

[22] Draggiotis P, Kleiss R H P and Papadopoulos C G 1998 On the computation of multigluon amplitudes *Phys. Lett.* B **439** 157–64

[23] Britto R, Cachazo F and Feng B 2005 New recursion relations for tree amplitudes of gluons *Nucl. Phys.* B **715** 499–522

[24] Britto R, Cachazo F, Feng B and Witten E 2005 Direct proof of tree-level recursion relation in Yang-Mills theory *Phys. Rev. Lett.* **94** 181602

[25] Mangano M L and Parke S J 1991 Multiparton amplitudes in gauge theories *Phys. Rep.* **200** 301–67

[26] Frixione S, Kunszt Z and Signer A 1996 Three jet cross-sections to next-to-leading order *Nucl. Phys.* B **467** 399–442

[27] Catani S and Seymour M H 1997 A General algorithm for calculating jet cross-sections in NLO QCD *Nucl. Phys.* B **485** 291–419 [Erratum: 1998 *Nucl. Phys.* B **510** 503–504]

[28] Harris B W and Owens J F 2002 The two cutoff phase space slicing method *Phys. Rev.* D **65** 094032

[29] Passarino G and Veltman M J G 1979 One loop corrections for $e^+ e^-$ annihilation into mu+ mu− in the Weinberg model *Nucl. Phys.* B **160** 151–207

[30] Ossola G, Papadopoulos C G and Pittau R 2007 Reducing full one-loop amplitudes to scalar integrals at the integrand level *Nucl. Phys.* B **763** 147–69

[31] Ellis R K, Kunszt Z, Melnikov K and Zanderighi G 2012 One-loop calculations in quantum field theory: from Feynman diagrams to unitarity cuts *Phys. Rep.* **518** 141–250

[32] Ellis R K, Giele W T, Kunszt Z, Melnikov K and Zanderighi G 2009 One-loop amplitudes for W^+ 3 jet production in hadron collisions *J. High Energy Phys.* **01** 012

[33] Andersen J R, Bartle S, Bern Z, Febres Cordero F, Höche S, Kosower D A, Ita H, Lo Presti N A, Maître D and Ozeren K 2014 High multiplicity processes with BlackHat and Sherpa *Proceedings of Loops and Legs in Quantum Field Theory—PoS(LL2014)* vol 211 (Trieste: SISSA)

[34] Cullen G *et al* 2014 GOSAM-2.0: a tool for automated one-loop calculations within the Standard Model and beyond *Eur. Phys. J.* C **74** 3001

[35] Bevilacqua G, Czakon M, Garzelli M V, van Hameren A, Kardos A, Papadopoulos C G, Pittau R and Worek M 2013 HELAC-NLO *Comput. Phys. Commun.* **184** 986–97

[36] Bern Z, Dixon L J, Febres Cordero F, Höche S, Ita H, Kosower D A, Maître D and Ozeren K J 2013 Next-to-leading order $W +$ 5-jet production at the LHC *Phys. Rev.* D **88** 014025

[37] Ita H, Bern Z, Dixon L J, Febres Cordero F, Kosower D A and Maitre D 2012 Precise predictions for $Z + 4$ jets at hadron colliders *Phys. Rev.* D **85** 031501

[38] Bern Z, Diana G, Dixon L J, Febres Cordero F, Hoeche S, Kosower D A, Ita H, Maitre D and Ozeren K 2012 Four-jet production at the large hadron collider at next-to-leading order in QCD *Phys. Rev. Lett.* **109** 042001

[39] Cullen G, van Deurzen H, Greiner N, Luisoni G, Mastrolia P, Mirabella E, Ossola G, Peraro T and Tramontano F 2013 Next-to-leading-order QCD corrections to Higgs boson production plus three jets in gluon fusion *Phys. Rev. Lett.* **111** 131801

[40] Greiner N, Höche S, Luisoni G, Schönherr M, Winter J-C and Yundin V 2016 Phenomenological analysis of Higgs boson production through gluon fusion in association with jets *J. High Energy Phys.* **01** 169

[41] Bevilacqua G, Czakon M, Papadopoulos C G, Pittau R and Worek M 2009 Assault on the NLO wishlist: pp —> t anti-t b anti-b *J. High Energy Phys.* **09** 109

[42] Denner A, Dittmaier S, Kallweit S and Pozzorini S 2012 NLO QCD corrections to off-shell top-antitop production with leptonic decays at hadron colliders *J. High Energy Phys.* **10** 110

[43] Buccioni F, Lang J-N, Lindert J M, Maierhöfer P, Pozzorini S, Zhang H and Zoller M F 2019 OpenLoops 2 *Eur. Phys. J.* C **79** 866

[44] Buccioni F, Kallweit S, Pozzorini S and Zoller M F 2019 NLO QCD predictions for $t\bar{t}b\bar{b}$ production in association with a light jet at the LHC *J. High Energy Phys.* **12** 015

[45] Nagy Z 2003 Next-to-leading order calculation of three jet observables in hadron-hadron collision *Phys. Rev.* D **68** 094002

[46] Caola F, Henn J M, Melnikov K and Smirnov V A 2014 Non-planar master integrals for the production of two off-shell vector bosons in collisions of massless partons *J. High Energy Phys.* **09** 043

[47] Gehrmann T, von Manteuffel A, Tancredi L and Weihs E 2014 The two-loop master integrals for $q\bar{q} \to VV$ *J. High Energy Phys.* **06** 32

[48] Bärnreuther P, Czakon M and Fiedler P 2014 Virtual amplitudes and threshold behaviour of hadronic top-quark pair-production cross sections *J. High Energy Phys.* **02** 078

[49] Anastasiou C, Gehrmann T, Oleari C, Remiddi E and Tausk J B 2000 The tensor reduction and master integrals of the two loop massless crossed box with lightlike legs *Nucl. Phys.* B **580** 577–601

[50] Chawdhry H A, Czakon M L, Mitov A and Poncelet R 2020 NNLO QCD corrections to three-photon production at the LHC *J. High Energy Phys.* **02** 057

[51] Czakon M, Mitov A and Poncelet R 2021 Next-to-next-to-leading order study of three-jet production at the LHC *Phys. Rev. Lett.* **127** 152001

[52] Gauld R, Gehrmann-De Ridder A, Glover E W N, Huss A and Majer I 2022 VH + jet production in hadron-hadron collisions up to order α_s^3 in perturbative QCD *J. High Energy Phys.* **03** 008

[53] Abreu S, Febres Cordero F, Ita H, Klinkert M, Page B and Sotnikov V 2022 Leading-color two-loop amplitudes for four partons and a W boson in QCD *J. High Energy Phys.* **04** 042

[54] Anastasiou C, Duhr C, Dulat F, Herzog F and Mistlberger B 2015 Higgs boson gluon-fusion production in QCD at three loops *Phys. Rev. Lett.* **114** 212001

[55] Duhr C, Dulat F and Mistlberger B 2020 Drell-Yan cross section to third order in the strong coupling constant *Phys. Rev. Lett.* **125** 172001

[56] Chen X, Gehrmann T, Glover E W N, Huss A, Mistlberger B and Pelloni A 2021 Fully differential Higgs boson production to third order in QCD *Phys. Rev. Lett.* **127** 072002
[57] Cacciari M, Dreyer F A, Karlberg A, Salam G P and Zanderighi G 2015 Fully differential vector-boson-fusion Higgs production at next-to-next-to-leading order *Phys. Rev. Lett.* **115** 082002 [Erratum: 2018 *Phys. Rev. Lett.* **120** 139901]
[58] Somogyi G, Trocsanyi Z and Del Duca V 2007 A subtraction scheme for computing QCD jet cross sections at NNLO: regularization of doubly-real emissions *J. High Energy Phys.* **01** 070
[59] Del Duca V, Duhr C, Somogyi G, Tramontano F and Trócsányi Z 2015 Higgs boson decay into b-quarks at NNLO accuracy *J. High Energy Phys.* **04** 036
[60] Del Duca V, Duhr C, Kardos A, Somogyi G, Szőr Z, Trócsányi Z and Tulipánt Z 2016 Jet production in the CoLoRFulNNLO method: event shapes in electron-positron collisions *Phys. Rev.* D **94** 074019
[61] Gehrmann-De Ridder A, Gehrmann T and Glover E W N 2005 Antenna subtraction at NNLO *J. High Energy Phys.* **09** 056
[62] Gehrmann-De Ridder A, Gehrmann T, Glover E W N and Heinrich G 2007 NNLO corrections to event shapes in e+ e− annihilation *J. High Energy Phys.* **12** 094
[63] Chen X, Cruz-Martinez J, Gehrmann T, Glover E W N and Jaquier M 2016 NNLO QCD corrections to Higgs boson production at large transverse momentum *J. High Energy Phys.* **10** 066
[64] Gehrmann-De Ridder A, Gehrmann T, Glover E W N, Huss A and Walker D M 2019 Vector boson production in association with a jet at forward rapidities *Eur. Phys. J.* C **79** 526
[65] Chen X, Gehrmann T, Glover E W N and Mo J 2022 Antenna subtraction for jet production observables in full colour at NNLO *J. High Energy Phys.* **10** 040
[66] Binoth T and Heinrich G 2000 An automatized algorithm to compute infrared divergent multiloop integrals *Nucl. Phys.* B **585** 741–59
[67] Anastasiou C, Melnikov K and Petriello F 2004 A new method for real radiation at NNLO *Phys. Rev.* D **69** 076010
[68] Anastasiou C, Melnikov K and Petriello F 2004 Higgs boson production at hadron colliders: differential cross sections through next-to-next-to-leading order *Phys. Rev. Lett.* **93** 262002
[69] Melnikov K and Petriello F 2006 Electroweak gauge boson production at hadron colliders through $O(\alpha_s^2)$ *Phys. Rev.* D **74** 114017
[70] Czakon M 2010 A novel subtraction scheme for double-real radiation at NNLO *Phys. Lett.* B **693** 259–68
[71] Czakon M, Fiedler P and Mitov A 2013 Total top-quark pair-production cross section at hadron colliders through $O(\alpha_S^4)$ *Phys. Rev. Lett.* **110** 252004
[72] Boughezal R, Caola F, Melnikov K, Petriello F and Schulze M 2015 Higgs boson production in association with a jet at next-to-next-to-leading order *Phys. Rev. Lett.* **115** 082003
[73] Catani S and Grazzini M 2007 An NNLO subtraction formalism in hadron collisions and its application to Higgs boson production at the LHC *Phys. Rev. Lett.* **98** 222002
[74] Catani S, Cieri L, Ferrera G, de Florian D and Grazzini M 2009 Vector boson production at hadron colliders: a fully exclusive QCD calculation at NNLO *Phys. Rev. Lett.* **103** 082001

[75] Ferrera G, Grazzini M and Tramontano F 2014 Higher-order QCD effects for associated WH production and decay at the LHC *J. High Energy Phys.* **04** 039

[76] Ferrera G, Grazzini M and Tramontano F 2015 Associated ZH production at hadron colliders: the fully differential NNLO QCD calculation *Phys. Lett.* B **740** 51–5

[77] Gehrmann T, Grazzini M, Kallweit S, Maierhöfer P, von Manteuffel A, Pozzorini S, Rathlev D and Tancredi L 2014 W^+W^- production at hadron colliders in next to next to leading order QCD *Phys. Rev. Lett.* **113** 212001

[78] Cascioli F, Gehrmann T, Grazzini M, Kallweit S, Maierhöfer P, von Manteuffel A, Pozzorini S, Rathlev D, Tancredi L and Weihs E 2014 ZZ production at hadron colliders in NNLO QCD *Phys. Lett.* B **735** 311–3

[79] Grazzini M, Kallweit S and Wiesemann M 2018 Fully differential NNLO computations with MATRIX *Eur. Phys. J.* C **78** 537

[80] Catani S, Devoto S, Grazzini M, Kallweit S and Mazzitelli J 2019 Top-quark pair production at the LHC: fully differential QCD predictions at NNLO *J. High Energy Phys.* **07** 100

[81] Boughezal R, Focke C, Giele W, Liu X and Petriello F 2015 Higgs boson production in association with a jet at NNLO using jettiness subtraction *Phys. Lett.* B **748** 5–8

[82] Gaunt J, Stahlhofen M, Tackmann F J and Walsh J R 2015 N-jettiness subtractions for NNLO QCD calculations *J. High Energy Phys.* **09** 058

[83] Ebert M A, Moult I, Stewart I W, Tackmann F J, Vita G and Zhu H X 2018 Power corrections for N-jettiness subtractions at $\mathcal{O}(\alpha_s)$ *J. High Energy Phys.* **12** 084

[84] Brown N and Stirling W J 1990 Jet cross-sections at leading double logarithm in e^+ e^- annihilation *Phys. Lett.* B **252** 657–62

[85] Catani S, Webber B R and Marchesini G 1991 QCD coherent branching and semiinclusive processes at large x *Nucl. Phys.* B **349** 635–54

[86] Bassetto A, Ciafaloni M and Marchesini G 1983 Jet structure and infrared sensitive quantities in perturbative QCD *Phys. Rep.* **100** 201–72

[87] Catani S, Trentadue L, Turnock G and Webber B R 1993 Resummation of large logarithms in e+ e− event shape distributions *Nucl. Phys.* B **407** 3–42

[88] Dasgupta M and Salam G P 2001 Resummation of nonglobal QCD observables *Phys. Lett.* B **512** 323–30

[89] Catani S, Turnock G and Webber B R 1992 Jet broadening measures in e^+e^- annihilation *Phys. Lett.* B **295** 269–76

[90] Dokshitzer Y L, Lucenti A, Marchesini G and Salam G P 1998 On the QCD analysis of jet broadening *J. High Energy Phys.* **01** 011

[91] Banfi A, Marchesini G and Smye G 2002 Away from jet energy flow *J. High Energy Phys.* **08** 006

[92] Dasgupta M and Salam G P 2002 Accounting for coherence in interjet E_t flow: a case study *J. High Energy Phys.* **03** 017

[93] Schwartz M D and Zhu H X 2014 Nonglobal logarithms at three loops, four loops, five loops, and beyond *Phys. Rev.* D **90** 065004

[94] Hagiwara Y, Hatta Y and Ueda T 2016 Hemisphere jet mass distribution at finite N_c *Phys. Lett.* B **756** 254–8

[95] Caron-Huot S 2018 Resummation of non-global logarithms and the BFKL equation *J. High Energy Phys.* **03** 036

[96] Becher T, Neubert M, Rothen L and Shao D Y 2016 Effective field theory for jet processes *Phys. Rev. Lett.* **116** 192001

[97] Banfi A, Dreyer F A and Monni P F 2021 Next-to-leading non-global logarithms in QCD *J. High Energy Phys.* **10** 006

[98] Banfi A, Dreyer F A and Monni P F 2022 Higher-order non-global logarithms from jet calculus *J. High Energy Phys.* **03** 135

[99] Becher T, Rauh T and Xu X 2022 Two-loop anomalous dimension for the resummation of non-global observables *J. High Energy Phys.* **08** 134

[100] Forshaw J R, Kyrieleis A and Seymour M H 2006 Super-leading logarithms in non-global observables in QCD *J. High Energy Phys.* **08** 059

[101] Becher T, Neubert M and Shao D Y 2021 Resummation of super-leading logarithms *Phys. Rev. Lett.* **127** 212002

[102] Bauer C W, Pirjol D and Stewart I W 2002 Soft collinear factorization in effective field theory *Phys. Rev.* D **65** 054022

[103] Becher T and Schwartz M D 2008 A precise determination of α_s from LEP thrust data using effective field theory *J. High Energy Phys.* **07** 034

[104] Stewart I W, Tackmann F J and Waalewijn W J 2010 N-jettiness: an inclusive event shape to veto jets *Phys. Rev. Lett.* **105** 092002

[105] Chien Y-T and Schwartz M D 2010 Resummation of heavy jet mass and comparison to LEP data *J. High Energy Phys.* **08** 058

[106] Becher T and Bell G 2012 NNLL resummation for jet broadening *J. High Energy Phys.* **11** 126

[107] Becher T and Neubert M 2012 Factorization and NNLL resummation for Higgs production with a jet veto *J. High Energy Phys.* **07** 108

[108] Lee R N, von Manteuffel A, Schabinger R M, Smirnov A V, Smirnov V A and Steinhauser M 2022 Quark and gluon form factors in four-loop QCD *Phys. Rev. Lett.* **128** 212002

[109] Magnea L 2001 Analytic resummation for the quark form-factor in QCD *Nucl. Phys.* B **593** 269–88

[110] von Manteuffel A, Panzer E and Schabinger R M 2020 Cusp and collinear anomalous dimensions in four-loop QCD from form factors *Phys. Rev. Lett.* **124** 162001

[111] Banfi A, Salam G P and Zanderighi G 2002 Semi-numerical resummation of event shapes *J. High Energy Phys.* **01** 018

[112] Banfi A, Salam G P and Zanderighi G 2005 Principles of general final-state resummation and automated implementation *J. High Energy Phys.* **03** 073

[113] Banfi A, McAslan H, Monni P F and Zanderighi G 2015 A general method for the resummation of event-shape distributions in e^+ e $^-$ annihilation *J. High Energy Phys.* **05** 102

[114] Banfi A, McAslan H, Monni P F and Zanderighi G 2016 The two-jet rate in e^+ e^- at next-to-next-to-leading-logarithmic order *Phys. Rev. Lett.* **117** 172001

[115] Banfi A, El-Menoufi B K and Monni P F 2019 The Sudakov radiator for jet observables and the soft physical coupling *J. High Energy Phys.* **01** 083

[116] Arpino L, Banfi A and El-Menoufi B K 2020 Near-to-planar three-jet events at NNLL accuracy *J. High Energy Phys.* **07** 171

[117] Gardi E and Magnea L 2009 Factorization constraints for soft anomalous dimensions in QCD scattering amplitudes *J. High Energy Phys.* **03** 079

[118] Bauer C W and Monni P F 2020 A formalism for the resummation of non-factorizable observables in SCET *J. High Energy Phys.* **05** 005

[119] Sjostrand T, Mrenna S and Skands P Z 2006 PYTHIA 6.4 physics and manual *J. High Energy Phys.* **05** 026

[120] Sjöstrand T, Ask S, Christiansen J R, Corke R, Desai N, Ilten P, Mrenna S, Prestel S, Rasmussen C O and Skands P Z 2015 An introduction to PYTHIA 8.2 *Comput. Phys. Commun.* **191** 159–77

[121] Sjostrand T and Skands P Z 2005 Transverse-momentum-ordered showers and interleaved multiple interactions *Eur. Phys. J.* C **39** 129–54

[122] Marchesini G and Webber B R 1984 Simulation of QCD jets including soft gluon interference *Nucl. Phys.* B **238** 1–29

[123] Marchesini G and Webber B R 1988 Monte Carlo simulation of general hard processes with coherent QCD radiation *Nucl. Phys.* B **310** 461–526

[124] Corcella G, Knowles I G, Marchesini G, Moretti S, Odagiri K, Richardson P, Seymour M H and Webber B R 2001 HERWIG 6: an event generator for hadron emission reactions with interfering gluons (including supersymmetric processes) *J. High Energy Phys.* **01** 010

[125] Bahr M *et al* 2008 Herwig++ physics and manual *Eur. Phys. J.* C **58** 639–707

[126] Gleisberg T, Hoeche S, Krauss F, Schalicke A, Schumann S and Winter J-C 2004 SHERPA 1. alpha: a proof of concept version *J. High Energy Phys.* **02** 056

[127] Banfi A, Corcella G and Dasgupta M 2007 Angular ordering and parton showers for non-global QCD observables *J. High Energy Phys.* **03** 050

[128] Lonnblad L 1992 ARIADNE version 4: a program for simulation of QCD cascades implementing the color dipole model *Comput. Phys. Commun.* **71** 15–31

[129] Ritzmann M, Kosower D A and Skands P 2013 Antenna showers with hadronic initial states *Phys. Lett.* B **718** 1345–50

[130] Höche S and Prestel S 2015 The midpoint between dipole and parton showers *Eur. Phys. J.* C **75** 461

[131] Dasgupta M, Dreyer F A, Hamilton K, Monni P F and Salam G P 2018 Logarithmic accuracy of parton showers: a fixed-order study *J. High Energy Phys.* **09** 033 [Erratum: 2020 J. High Energy Phys. 03 083]

[132] Dasgupta M, Dreyer F A, Hamilton K, Monni P F, Salam G P and Soyez G 2020 Parton showers beyond leading logarithmic accuracy *Phys. Rev. Lett.* **125** 052002

[133] van Beekveld M, Ferrario Ravasio S, Salam G P, Soto-Ontoso A, Soyez G and Verheyen R 2022 PanScales parton showers for hadron collisions: formulation and fixed-order studies *J. High Energy Phys.* **2022** 19

[134] Nagy Z and Soper D E 2014 A parton shower based on factorization of the quantum density matrix *J. High Energy Phys.* **06** 097

[135] Forshaw J R, Holguin J and Plätzer S 2019 Parton branching at amplitude level *J. High Energy Phys.* **08** 145

[136] Catani S, Krauss F, Kuhn R and Webber B R 2001 QCD matrix elements + parton showers *J. High Energy Phys.* **11** 063

[137] Mangano M L, Moretti M and Pittau R 2002 Multijet matrix elements and shower evolution in hadronic collisions: $Wb\bar{b} + n$ jets as a case study *Nucl. Phys.* B **632** 343–62

[138] Frixione S and Webber B R 2002 Matching NLO QCD computations and parton shower simulations *J. High Energy Phys.* **06** 029

[139] Nason P 2004 A new method for combining NLO QCD with shower Monte Carlo algorithms *J. High Energy Phys.* **11** 040
[140] Hoeche S, Krauss F, Schumann S and Siegert F 2009 QCD matrix elements and truncated showers *J. High Energy Phys.* **05** 053
[141] Frederix R and Frixione S 2012 Merging meets matching in MC@NLO *J. High Energy Phys.* **12** 061
[142] Aad G *et al* (ATLAS Collaboration) 2013 Measurement of the production cross section of jets in association with a Z boson in pp collisions at $\sqrt{s} = 7$ TeV with the ATLAS detector *J. High Energy Phys.* **07** 032
[143] Sjostrand T, Mrenna S and Skands P Z 2008 A Brief introduction to PYTHIA 8.1 *Comput. Phys. Commun.* **178** 852–67
[144] Lonnblad L 2002 Correcting the color dipole cascade model with fixed order matrix elements *J. High Energy Phys.* **05** 046
[145] Bothmann E *et al* (Sherpa Collaboration) 2019 Event generation with Sherpa 2.2 *SciPost Phys.* **7** 034
[146] Mangano M L, Moretti M, Piccinini F, Pittau R and Polosa A D 2003 ALPGEN, a generator for hard multiparton processes in hadronic collisions *J. High Energy Phys.* **07** 001
[147] Boughezal R, Liu X and Petriello F 2016 Phenomenology of the Z-boson plus jet process at NNLO *Phys. Rev.* D **94** 074015
[148] ATLAS Collaboration 2022 Cross-section measurements for the production of a Z boson in association with high-transverse-momentum jets in pp collisions at $\sqrt{s}$ TeV with the ATLAS detector
[149] Gehrmann-De Ridder A, Gehrmann T, Glover E W N, Huss A and Walker D M 2018 Next-to-next-to-leading-order QCD corrections to the transverse momentum distribution of weak gauge bosons *Phys. Rev. Lett.* **120** 122001
[150] Denner A, Dittmaier S, Kasprzik T and Muck A 2011 Electroweak corrections to dilepton + jet production at hadron colliders *J. High Energy Phys.* **06** 069
[151] Aad G *et al* (ATLAS Collaboration) 2020 Measurement of the transverse momentum distribution of Drell–Yan lepton pairs in proton–proton collisions at $\sqrt{s} = 13$ TeV with the ATLAS detector *Eur. Phys. J.* C **80** 616
[152] Chen X, Gehrmann T, Glover E W N, Huss A, Monni P F, Re E, Rottoli L and Torrielli P 2022 Third-order fiducial predictions for Drell-Yan production at the LHC *Phys. Rev. Lett.* **128** 252001
[153] Monni P F, Re E and Torrielli P 2016 Higgs transverse-momentum resummation in direct space *Phys. Rev. Lett.* **116** 242001
[154] Hamilton K, Nason P, Oleari C and Zanderighi G 2013 Merging H/W/Z + 0 and 1 jet at NLO with no merging scale: a path to parton shower + NNLO matching *J. High Energy Phys.* **05** 082
[155] Gehrmann T, Hoche S, Krauss F, Schonherr M and Siegert F 2013 NLO QCD matrix elements + parton showers in $e^+ e^-$ -> hadrons *J. High Energy Phys.* **01** 144
[156] Alioli S, Bauer C W, Berggren C J, Hornig A, Tackmann F J, Vermilion C K, Walsh J R and Zuberi S 2013 Combining higher-order resummation with multiple NLO calculations and parton showers in GENEVA *J. High Energy Phys.* **09** 120
[157] Hamilton K, Nason P, Re E and Zanderighi G 2013 NNLOPS simulation of Higgs boson production *J. High Energy Phys.* **10** 222

[158] Buonocore L *et al* 2022 NNLO+PS with MiNNLO$_{\rm PS}$: status and prospects arXiv:2203.07240 [hep-ph]

[159] Brock R *et al* (CTEQ Collaboration) 1995 Handbook of perturbative QCD: version 1.0 *Rev. Mod. Phys.* **67** 157–248

[160] Workman R L *et al* (Particle Data Group Collaboration) 2022 Review of particle physics *Prog. Theor. Exp. Phys.* **2022** 083C01

[161] Shifman M A 2001 Quark-hadron duality *At the Frontier of Particle Physics: Handbook of QCD* (Singapore: World Scientific) pp 1447–94

[162] Dokshitzer Y L, Khoze V A, Mueller A H and Troian S I 1991 *Basics of Perturbative QCD* (Singapore: Editions Frontiéres)

[163] Andersson B, Gustafson G, Ingelman G and Sjostrand T 1983 Parton fragmentation and string dynamics *Phys. Rep.* **97** 31–145

[164] Dokshitzer Y L, Marchesini G and Webber B R 1996 Dispersive approach to power behaved contributions in QCD hard processes *Nucl. Phys.* B **469** 93–142

[165] Dokshitzer Y L and Webber B R 1997 Power corrections to event shape distributions *Phys. Lett.* B **404** 321–7

[166] Korchemsky G P and Sterman G F 1999 Power corrections to event shapes and factorization *Nucl. Phys.* B **555** 335–51

[167] Heister A *et al* (ALEPH Collaboration) 2004 Studies of QCD at $e^+ e^-$ centre-of-mass energies between 91-GeV and 209-GeV *Eur. Phys. J.* C **35** 457–86

[168] Achard P *et al* (L3 Collaboration) 2004 Studies of hadronic event structure in e^+e^- annihilation from 30-GeV to 209-GeV with the L3 detector *Phys. Rep.* **399** 71–174

[169] Dokshitzer Y L, Lucenti A, Marchesini G and Salam G P 1998 On the universality of the Milan factor for $1/Q$ power corrections to jet shapes *J. High Energy Phys.* **05** 003

[170] Caola F, Ferrario Ravasio S, Limatola G, Melnikov K, Nason P and Ozcelik M A 2022 Linear power corrections to e^+e^- shape variables in the three-jet region *J. High Energ. Phys.* **2022** 62

[171] Bartalini P and Gaunt J R (ed) 2019 *Multiple Parton Interactions at the LHC* vol 29 (Singapore: World Scientific)

[172] Sjöstrand T 2018 The development of MPI modelling in PYTHIA *Adv. Ser. Direct. High Energy Phys.* **29** 191–225

[173] Butterworth J M, Forshaw J R and Seymour M H 1966 Multiparton interactions in photoproduction at HERA *Z. Phys.* C **72** 637–46

[174] Field R D (CDF Collaboration) 2001 The underlying event in hard scattering processes *eConf* **C010630** P501 [arXiv:hep-ph/0201192]

[175] Cacciari M, Salam G P and Sapeta S 2010 On the characterisation of the underlying event *J. High Energy Phys.* **04** 065

IOP Publishing

Hadronic Jets (Second Edition)
An introduction
Andrea Banfi

Chapter 4

Jets as discovery tools

As we have seen in the previous chapter, quantum chromodynamics (QCD), the theory of strong interactions, predicts the occurrence of jets in high-energy collisions. Using perturbative QCD it is possible to describe observables involving jets with high theoretical accuracy. A perturbative description of jet observables is particularly appropriate at very high scales, since non-perturbative corrections (e.g. hadronisation) are suppressed by inverse powers of the characteristic energy of each process. Therefore, the most natural use of jets is that of testing perturbative QCD, for instance by measuring the QCD running coupling. One such measurement is the inclusive transverse momentum distribution of a jet in hadron collisions, obtained by binning the transverse momentum of every detected jet. The plot in the left-hand panel of figure 4.1 shows the most recent of such measurements, performed by the ATLAS collaboration at the LHC [1]. One sees excellent agreement between the data and NLO QCD predictions obtained with the program NLOJET++ [2], so that an extraction of the QCD coupling α_s is possible (see e.g. [3] for an α_s determination by CMS with older data). Even more precise NNLO predictions obtained with the program NNLOJET [4] are now available. These are currently used for fits of parton distribution functions [5, 6]. Inclusive jet transverse momentum spectra are just an example of the many jet observables that can be used for α_s measurements. A comprehensive plot showing the summary of the most precise measurements of α_s [7] is shown on the right-hand panel of figure 4.1. There is a striking agreement between all the measurements and the running of the coupling predicted by QCD, thus confirming its reliability as a theory of strong interactions at high energies. Note also that not only do jet observables provide most of the measurements in figure 4.1 (jets in electron–positron collisions, e^+e^- jets and shapes, pp, $p\bar{p} \to$ jets), but also make it possible to explore different energy scales, which is not possible with fully inclusive observables such as τ decays, who are characterised by a single hard scale. This is why it is very important to refine theoretical methods as much as possible to compute jet observables with higher and higher precision.

Using jets as a means of testing perturbative QCD is just one of the ways in which they can be exploited. Another important application of jet physics is the search for

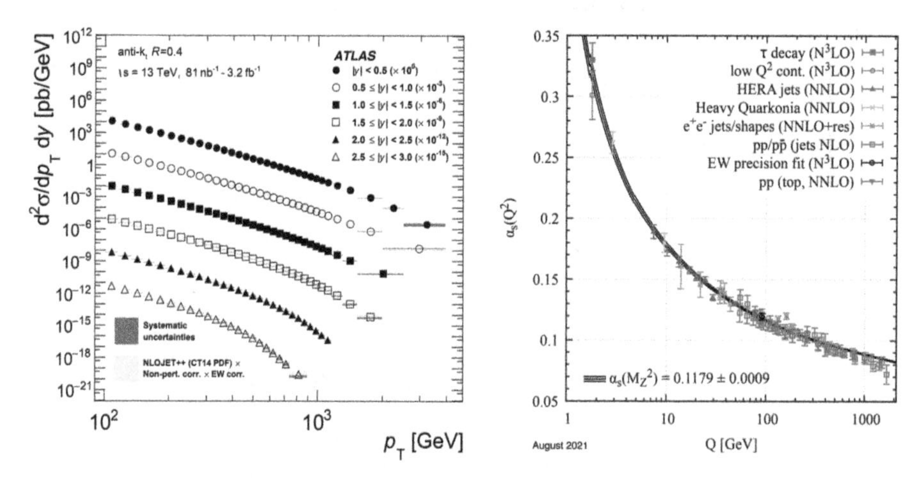

Figure 4.1. The inclusive transverse momentum spectra of jets, as measured by the ATLAS collaboration in different rapidity bins (left, © [1] (2018), with permission of Springer), and the running of the QCD coupling, as determined by various measurements (right, reproduced from [7], copyright © 2022, Oxford University Press).

new particles, especially at hadron colliders. To understand how this can be achieved, let us start again with an example, taken from [8]. Suppose we are looking for a new heavy particle that decays into quarks. The latter will give rise to jets in the final state. The relevant question is how to distinguish jets originating from new-particle decays from those produced directly in a hard collision via strong interactions. The main strategy is to devise suitable observables whose distributions are different for signal and background events, thus acting as discriminants. The choice of these observables may be guided by a trial-error procedure, e.g. simulating signal and backgrounds with parton-shower event generators and examining the outcome until a satisfactory signal-to-background ratio is obtained. This approach is reasonable, and highlights how important it is that parton-shower event generators accurately describe most of the features of jet production. However, this information is usefully complemented by theoretical considerations. For instance, consider the simple scenario in which one looks for a heavy electrically neutral vector boson, let's call it Z', that decays into a quark–antiquark pair. The natural way to search for such an object at the LHC is to reconstruct a pair of jets at high transverse momentum, say with the anti-k_t algorithm with a given radius R, and look for a peak in the invariant mass distribution of the two jets. The choice of the jet radius plays an important role in such analyses. In fact, partons from Z' decay can radiate outside the jet radius, thus causing the dijet pair to loose mass, and instead of a peak we may observe a broad distribution, indistinguishable from the continuum background. If, on the contrary, we increase the radius too much, we catch many secondary hadrons from pile-up (PU) and from the underlying event (UE), again spoiling the resolution of the mass peak. This effect can be appreciated in figure 4.2, where (simulated) jets originating from the decay of a Z' with a mass of 2 TeV are reconstructed with the anti-k_t algorithm and different jet radii. There one clearly sees a degradation of the reconstruction quality of the mass peak when one moves away

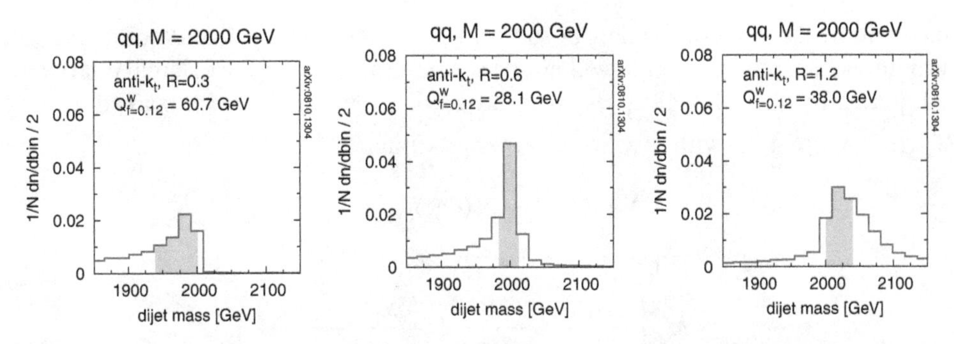

Figure 4.2. The distribution in the invariant mass of two quark jets in hadron collisions with UE, but no PU, obtained from [8]. © Matteo Cacciari, Juan Rojo and Gavin P Salam.

from $R = 0.6$, which represents somehow an optimal radius for this analysis. A trial-error procedure in this context consists in reconstructing the mass peak for all possible algorithms and jet values, and this would have to be repeated for each new analysis. Furthermore, relying on the outcome of parton-shower event generators implicitly assumes that these tools are able to capture the main features of the physics involved. A theoretically more sound approach consists in assessing the impact of perturbative and non-perturbative effects on jet-based analyses, and checking whether parton-shower event generators reproduce the main features that the theory predicts. The thus-validated event generators can be reliably used in a more refined trial-error procedure, for instance to train a neural network. In this chapter we will discuss how QCD has helped in devising better procedures to discriminate signal events, e.g. the hadronic decay of a heavy particle, from background events arising from QCD jet production. In particular, in section 4.1 we will discuss how the energy-momentum content of a jet of a given radius is affected by QCD radiation and non-perturbative effects like hadronisation and UE. In section 4.2 we will present an overview of searches for boosted particles, whose decay products tend to fall into the same jet. Finally, in section 4.3 we will discuss methods to distinguish whether a jet is initiated by a quark or a gluon, highlighting why this is important for new physics searches at colliders.

4.1 Optimising the jet radius

Consider again the case of a Z' produced in hadron–hadron collisions and decaying into two jets (identified by their transverse momenta p_{t1}, p_{t2}, rapidities y_1, y_2 and azimuthal angles ϕ_1, ϕ_2, all with respect to the beam), where we look for a peak in the dijet invariant mass distribution $d\sigma/dM_{jj}^2$. The invariant mass M_{jj} is related to the jet kinematic variables through the relation

$$M_{jj}^2 \simeq M_1^2 + M_2^2 + 2p_{t1}p_{t2}\left[\cosh(y_1 - y_2) - \cos(\phi_1 - \phi_2)\right] \qquad (4.1)$$

where we have introduced the invariant masses M_1, M_2 of the two jets. These are in general acquired dynamically through QCD radiation, or by collecting particles from UE/PU. If p_1 and p_2 are just a quark and an antiquark, their total invariant mass will be the mass of the new boson $M_{Z'}$. However, the quark and antiquark will turn into jets, whose momenta will not be the same as those of the original quark and antiquark. In turn, this gives rise to a broad distribution $d\sigma/dM_{jj}^2$, peaked around $M_{Z'}^2$ (see figure 4.2), with a width given approximately by

$$\begin{aligned}\delta M_{jj}^2 &\simeq \langle M_1^2\rangle + \langle M_2^2\rangle + 2(\langle \delta p_{t1}\rangle p_{t2} + p_{t1}\langle \delta p_{t2}\rangle)\\ &\quad \left[\cosh(y_1 - y_2) - \cos(\phi_1 - \phi_2)\right]\\ &= M_{Z'}^2\left(\frac{\langle M_1^2\rangle}{M_{Z'}^2} + \frac{\langle M_2^2\rangle}{M_{Z'}^2} + \frac{\langle \delta p_{t1}\rangle}{p_{t1}} + \frac{\langle \delta p_{t2}\rangle}{p_{t2}}\right),\end{aligned} \tag{4.2}$$

where $\langle \delta p_t\rangle$ is difference between the transverse momentum of a jet p_t and that of the parton that has initiated it, averaged over all possible final-state configurations. Similarly, $\langle M^2\rangle$ represents the average invariant mass squared of a jet. Here we have assumed that jet directions stay approximately unchanged. There are three main effects contributing to $\langle \delta p_t\rangle$ and to the invariant mass of a jet: QCD radiation, hadronisation, and PU/UE. Here we will concentrate on the change in transverse momentum and present the main features of each contribution. We will also present the main results for the invariant mass. For a more detailed analysis, the reader is referred to the original source [9].

QCD radiation. Consider one of the two jets produced by the decay of a Z' into a quark–antiquark pair, for instance the one initiated by the quark. When a gluon is emitted from the quark, it can escape the jet, so that the jet transverse momentum after the splitting is less than the quark transverse momentum. Suppose also that our candidate jet is the one with the highest transverse momentum, which is reasonable because we want to avoid following soft jets to search for new physics. In quasi-collinear kinematics, the transverse momentum of the hardest jet after the splitting will be $\max(z, 1-z)p_t$, with p_t the transverse momentum of the parent quark and z the splitting fraction (see section 3.2). Therefore, the quantity $\delta p_t = -\min(z, 1-z)p_t$ represents the transverse momentum lost in the splitting. The average decrease $\langle \delta p_t\rangle$ is obtained by integrating δp_t over the phase space of the emitted gluon, with the condition that the latter is not clustered with the outgoing quark. This condition depends in general on the jet algorithm. However, for generalised k_t-algorithms, a gluon escapes a jet if and only if its distance ΔR in the y–ϕ plane from the final-state quark is bigger than the jet radius R, as illustrated in figure 4.3. For small angles, ΔR

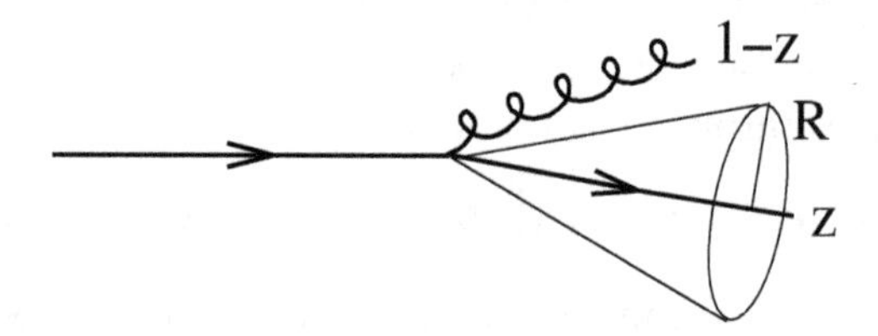

Figure 4.3. The collinear emission of a gluon, carrying an energy fraction $1 - z$ of the parent quark. The gluon is softer than the final-state quark, and is not clustered with it.

is approximately equal to the splitting angle θ, so that the corresponding transverse momentum loss is given by

$$\begin{aligned}\langle \delta p_t \rangle_{q,\mathrm{PT}} &\simeq \frac{\alpha_s(p_t)}{2\pi} \int_{R^2}^{1} \frac{d\theta^2}{\theta^2} \int_0^1 dz \; P_{q\to qg}(z)\big[-\min(z,\, 1-z)p_t\big] \\ &= -\, C_F\left(2\ln 2 - \frac{3}{8}\right) p_t \frac{\alpha_s}{\pi} \ln\frac{1}{R} \simeq -1.35\, p_t \frac{\alpha_s(p_t)}{\pi} \ln\frac{1}{R},\end{aligned} \tag{4.3}$$

where $P_{q\to qg}(z)$ is the splitting function related to the elementary process $q \to qg$, defined in equation (3.33). Given the fact that the splitting probability is positive, we have a transverse momentum loss, as expected. Notice that this quantity is logarithmically enhanced in the jet radius, a consequence of the collinear singularity $d\theta^2/\theta^2$ in the gluon emission probability. This means that the smaller the jet radius, the more the jet will lose transverse momentum due to QCD radiation, hence spoiling the resolution of an invariant mass peak.

For jets initiated by gluons, one needs to consider a gluon splitting into two gluons or n_f quark–antiquark pairs, with a mass that is negligible with respect to the jet transverse momentum (see figure 4.4). Using the expressions for $P_{g\to gg}(z)$ and $P_{g\to q\bar{q}}$ in equation (3.33), we obtain

$$\begin{aligned}\langle \delta p_t \rangle_{g,\mathrm{PT}} &\simeq \frac{\alpha_s}{2\pi} \int_{R^2}^{1} \frac{d\theta^2}{\theta^2} \int_0^1 dz \, \big[P_{g\to gg}(z) + n_f P_{g\to q\bar{q}}(z)\big] \\ &\qquad \big[-\min(z,\, 1-z)p_t\big] \\ &= -\left(\left(2\ln 2 - \frac{43}{96}\right) C_A + \frac{7}{48} T_F n_f\right) p_t \frac{\alpha_s}{\pi} \ln\frac{1}{R} \\ &\simeq -\,(2.82 + 0.07 n_f)\, p_t \frac{\alpha_s}{\pi} \ln\frac{1}{R}.\end{aligned} \tag{4.4}$$

By comparing the size of the two contributions one sees that gluon-initiated jets lose roughly twice as much transverse momentum with respect to quark-initiated jets. In fact, the ratio $\langle \delta p_t \rangle_g / \langle \delta p_t \rangle_q$ is driven by the ratio of the corresponding Casimir factors C_A/C_F, as the corresponding splitting probabilities are dominated by the universal soft singularity $dz/(1-z)$.

Note that, for small jet radii, the logarithms of the jet radius in equations (4.3) and (4.4) can become large, thus endangering the convergence of the QCD perturbative expansion. Logarithms of the jet radius can be resummed at all orders

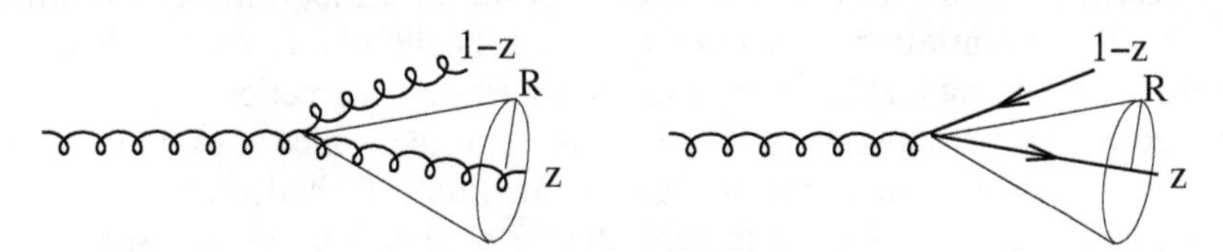

Figure 4.4. A gluon splitting into either two gluons or a quark–antiquark pair, where the two partons are not clustered in the same jet.

by exploiting the coherent branching formalism, and considering a cascade of collinear splittings with decreasing angles, which stops when the angle of the most collinear splitting reaches the jet radius R. Such resummation has been carried out in [10].

In the case of the jet invariant mass, we obtain a non-zero contribution when the emitted gluon is clustered within the jet. This gives $\langle M^2 \rangle_{\mathrm{PT}} \sim \alpha_s p_t^2 R^2$ [11], so smaller jet radii give a smaller perturbatively generated jet mass.

Hadronisation. Suppose some soft hadrons at large angles escaped the jet. As explained in section 3.3, it is useful, as well as phenomenologically accurate, to treat such hadrons as ultra-soft gluons emitted with an effective coupling $\alpha_s^{\mathrm{NP}}(k_t)$, where k_t is the relative transverse momentum of the emitted non-perturbative gluon with respect to the emitter. For an ultra-soft gluon taking a fraction $1 - z$ of jet transverse momentum, we have $k_t \simeq (1 - z)\theta p_t$. The loss of these soft hadrons will not change appreciably the transverse momentum of the leading jet, so that the transverse momentum loss due to hadronisation is always $\delta p_t \simeq -(1 - z)p_t$. This elementary process can be visualised again in figure 4.3, this time with $z \simeq 1$. This gives, for a quark-initiated jet

$$\begin{aligned}\langle \delta p_t \rangle_{q,\mathrm{had}} &\simeq \int_{R^2}^{1} \frac{d\theta^2}{\theta^2} \int_0^1 dz \frac{2C_F}{1-z}\left[-(1-z)p_t\right] \frac{\alpha_s^{\mathrm{NP}}[(1-z)\theta p_t]}{2\pi} \\ &\simeq -\frac{2C_F}{\pi} \int_R^1 \frac{d\theta}{\theta^2} \int_0^{\mu_I} dk_t\, \alpha_s^{\mathrm{NP}}(k_t).\end{aligned} \tag{4.5}$$

Note that we are entitled to use the most singular part (for $z \to 1$) of the splitting function only, because the remaining parts give a correction that is suppressed by a higher power of the jet transverse momentum. We have also bounded the relative transverse momentum k_t at a scale $\mu_I \simeq 2$ GeV as a boundary between perturbative and non-perturbative physics, see section 3.3. Using equation (3.86), the average transverse momentum loss can be further expressed in terms of the same parameter $\alpha_0(\mu_I)$ that enters hadronisation corrections to event-shape distributions and means in e^+e^- annihilation, as follows

$$\frac{\langle \delta p_t \rangle_{q,\mathrm{had}}}{p_t} \simeq -\frac{8C_F}{\pi^2} \frac{1}{R} \frac{\mu_I}{p_t}\left[\alpha_0(\mu_I) - \alpha_s(p_t) + \mathcal{O}(\alpha_s^2)\right]. \tag{4.6}$$

For a gluon jet the answer is almost identical, as one needs to replace C_F with C_A. Since $C_A \simeq 2C_F$, it turns out that a gluon jet loses on average twice as much transverse momentum with respect to a quark jet through hadronisation. Notice also that the relative p_t-loss scales like one inverse power of the jet transverse momentum. As expected, hadronisation corrections become negligible at very high transverse momenta. The leading $1/R$ behaviour of hadronisation corrections is reproduced by parton-shower event generators [9]. This is expected, since a similar correction appears in event-shape distributions and means, and all hadronisation models in event generators are tuned so as to reproduce e^+e^- data for event shapes.

Hadronisation corrections to the jet mass are obtained in a similar way. This time, the emitted ultra-soft gluon has to be clustered within the jet to give a non-zero contribution to the jet mass. This gives a correction $\langle M^2 \rangle_{\rm had} \sim \mu_I \alpha_0(\mu_I) p_t R$ [9], i.e. increasing with the jet radius.

Pile-up and/or underlying event. The common characteristics of the p_t-loss induced by QCD radiation and hadronisation is that they increase with decreasing radius. However, in hadron colliders, a smaller p_t-loss cannot be obtained by just increasing the jet radius, because of contamination from pile-up and underlying event. These contributions can be discussed together in that they produce a distribution in transverse momentum which is roughly uniform in rapidity and azimuth. Note that, in this context, a further contamination with the same characteristics is given by soft, perturbative or non-perturbative, radiation from the initial-state partons. In our discussion then we will simply consider a generic source of background, producing a uniform average transverse momentum $\rho_{\rm bkg}$ per unit rapidity and unit azimuth. It follows immediately that such background produces an average increase $\langle \delta p_t \rangle_{\rm bkg}$ of the transverse momentum of a jet that gets larger with the jet radius. Assuming E-scheme recombination for the jets, the change in transverse momentum δp_t due to background hadrons having a total transverse momentum $\vec{k}_t$ can be estimated as follows

$$\delta p_t = |\vec{p}_t + \vec{k}_t| - p_t \simeq \frac{\vec{k}_t \cdot \vec{p}_t}{p_t} \simeq k_t \cos\phi, \tag{4.7}$$

with ϕ the azimuthal angle between $\vec{p}_t$ and $\vec{k}_t$. If we are interested only in the leading dependence on the jet radius, we can take the limit of small R, and hence $\cos\phi \simeq 1$, so that $\delta p_t \simeq k_t$. This situation is depicted in figure 4.5. If we average over all possible values of k_t, and integrate over the region in rapidity and azimuth corresponding to the area of the jet, we obtain [9]

$$\langle \delta p_t \rangle_{\rm bkg} \simeq \pi R^2 \rho_{\rm bkg}$$

with πR^2 the active area of an anti-k_t jet. Notice that, for small R, the same result holds for the Cambridge/Aachen and the k_t algorithms. We recall that the contamination due to a uniform background can be removed by means of the subtraction procedure of equation (2.7). To apply that equation one needs to find an estimator of ρ_{bkg} itself using data, for instance the one defined in equation (2.10).

A similar calculation for the jet mass gives a contamination due to a uniform background $\langle M^2 \rangle_{\rm bkg} \sim \rho_{\rm bkg} p_t R^4$, increasing with the jet radius.

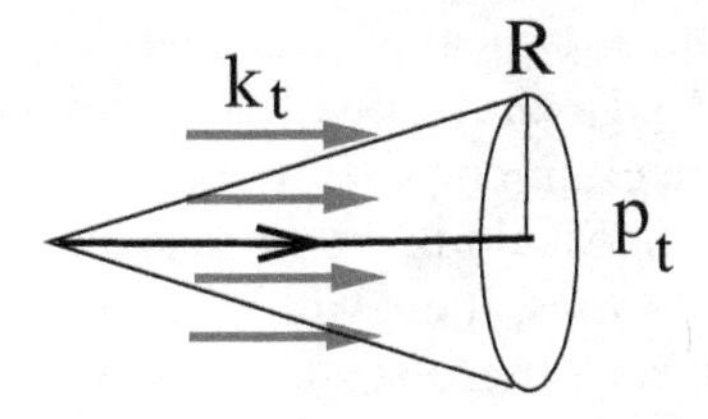

Figure 4.5. Particles from a uniform background (the red arrows, each with its transverse momentum k_t) entering a quark jet (transverse momentum p_t, radius R) quasi parallel to the jet direction.

Concluding remarks. The calculations we have presented so far convey some important messages. The dominant effect of QCD final-state radiation (perturbative or non-perturbative) is that of degrading the transverse momentum of a jet, the larger the decrease the smaller the jet radius, and of increasing the jet invariant mass with the jet radius. In contrast, sources of a uniform background in rapidity and azimuth, like initial-state QCD radiation, pile-up or underlying event, cause an increase both in a jet transverse momentum and in its invariant mass, increasing with increasing jet radius. One can exploit this information to find an optimal jet radius, for which the combination of all effects is minimised (see e.g. [9] for fully worked-out examples). The way the minimisation is performed depends crucially on the analysis one is interested in. It is clear that it is desirable to keep the jet radius small in order to minimise contamination from a uniform background. However, if the jet radius is too small, perturbative effects and hadronisation become important. To be concrete, if the focus is on the accurate description of the transverse momentum distribution of a jet, one can consider only the change in transverse momentum induced by hadronisation and uniform background. If instead one is interested in new physics searches, where one wants to preserve as much information as possible on the partons that have initiated each jet, it is crucial to minimise effects due to perturbative radiation as well. Although strictly speaking this is not needed for perturbative calculations, perturbative effects that are proportional to $\ln(1/R)$ can induce large higher-order corrections and then spoil the convergence of perturbative expansions. Therefore, reducing the transverse momentum loss due to perturbative radiation might help reduce theoretical uncertainties of precision calculations. Another useful observation is that, since the effect of QCD radiation is proportional to the colour charge of the parton initiating the jet, whereas the contribution of the UE is independent of it, one expects that the optimal radius will be roughly twice as large for gluon jets than for quark jets.

To conclude, finding the optimal jet radius is one of the crucial problems when using jets both in precision studies and in new physics searches. Here we have presented only basic observations based on perturbative QCD and models of hadronisation and UE/PU, which do not by any means substitute a dedicated study according to the detailed experimental analyses of interest. One of such experimental investigations, namely the search for boosted objects, whose decay products fall into the same jet, deserve a special mention. This will be covered in the next section.

4.2 Boosted objects and jet substructure

The LHC makes it possible for the first time to access in a controlled way energy scales that are larger than those typical of electroweak interactions, which are of the order of a few hundreds of giga-electron-volts. In this situation, a massive particle can have a transverse momentum considerably larger than its mass, and its decay products will receive a huge boost along the particle direction. As a consequence, if a heavy particle decays into partons, these will be likely clustered inside the same jet. In fact, for a jet of mass m_{jet}, the distance ΔR_{12} in the rapidity–azimuth plane between two quasi-parallel constituents, carrying a fraction z and $1 - z$ of the jet transverse momentum $p_{t,\text{jet}}$, is given by

$$\Delta R_{12} \simeq \frac{m_{\text{jet}}}{\sqrt{z(1-z)}p_{\text{t,jet}}}, \tag{4.8}$$

which, for fixed m_{jet}, becomes smaller the larger the jet transverse momentum. Boosted analyses consider then a candidate jet, and look for a peak in the distribution in the invariant mass of that jet. Various problems arise in this situation. The first is how to distinguish such jets from standard QCD jets, whose invariant mass is dynamically generated through parton branching. The peak of the mass distribution of a QCD jet is at around 10% of its transverse momentum. Therefore, for a jet of the transverse momentum of 1 TeV, we expect a peak at around 100 GeV, right at the electroweak scale. The other problem is to clean the jet in such a way that only the hadrons originated by the decay of the heavy particle are used to compute the jet invariant mass. Here similar considerations apply as for the mass resolution of dijet pairs, discussed in the previous section. Suppose for instance that a new particle decays into two partons. These lose energy due to QCD radiation, so it might be possible that some of these escape the jet, resulting in a lower invariant mass than expected. One could then enlarge the jet radius R, but we have seen that contamination of the mass of a jet from a uniform background scales like R^4. Therefore, one needs to find an optimal procedure to clean the candidate jet down to its relevant constituents.

The methods used to search for jets arising from the decay of heavy particles are generally called 'jet substructure techniques'. They were introduced for the first time in [12], in the context of searches for a heavy Higgs boson decaying into a pair of W bosons, one of which decays hadronically. The basic idea of that analysis was to consider a jet with a large radius, which we now call a 'fat' jet, measure the invariant mass of its two hardest subjets and look for a peak corresponding to the mass of the W boson. To reduce contamination from the underlying event it was also proposed to recluster hadrons inside the fat jet with a smaller jet radius, whose size was optimised according to the desired resolution in the fat-jet invariant mass. In the following we give an overview of the most used methods to search for boosted heavy objects inside fat jets. For simplicity, we will concentrate on two-prong decays, in which a heavy particle decays into two coloured particles, referring the reader to reviews on the subject, such as [13], for the generalisation to three-prong decays (for relevant examples of top-quark taggers see e.g. [14–17]).

The most widely used procedures for jet-substructure studies perform two different kinds of actions on a jet. One is grooming, aimed at cleaning a jet from soft constituents, whose output is always a jet, which might be very different from the original jet. These soft constituents can have any origin, can be for instance soft gluons radiated by the hard partons that have initiated the jet, as well as contamination from initial-state radiation, PU or UE. The second action is tagging, through which a jet can be either kept or discarded, according to whether it satisfies certain criteria, specific to the particle we are looking for. Note that grooming and tagging cannot always be separated, as sometimes a groomer acts as a tagger. This happens for instance when a groomer leaves two well-separated subjets, which are good candidates for a two-prong decay.

4.2.1 Groomers

In this section we outline the basic principles of jet grooming. We will do this by analysing the widely used grooming strategies of trimming [18] and pruning [19]. We will then consider the mass-drop tagger (MDT) [20], which incorporates both grooming and tagging procedures. In section 4.2.1 we will discuss only one of the grooming procedures of the MDT, leaving the tagging for section 4.2.2. We will conclude with the soft drop [21] in grooming mode, which generalises the MDT and is one of the most popular grooming procedures.

Trimming. This procedure aims at cleaning all hard jets (i.e. those above some transverse momentum threshold) from their softer constituents [18]. In practice, once jets have been reconstructed, each jet is reclustered with a radius R_{sub}, smaller than the jet radius. One then considers each jet in turn and discards all subjets p_j having a transverse momentum $p_{tj} < z_{\text{cut}}\Lambda_{\text{hard}}$, where z_{cut} is a free parameter, and Λ_{hard} a typical hard scale, for instance the transverse momentum of the original jet. The surviving subjets constitute the trimmed jet. Figure 4.6 illustrates pictorially how the trimming algorithm works with a jet containing two hard and two soft constituents. In this specific example, at the end of the procedure one obtains a hard jet of radius R_{sub} containing the two hard constituents only.

Pruning. This procedure aims at eliminating soft large-angle constituents of jets by trying to follow the hardest branch of a jet [19]. This is achieved by performing a reclustering of the constituents of each jet with a sequential algorithms such that, whenever two constituents p_i and p_j are to be recombined, one checks whether they meet the conditions

$$\frac{\min(p_{ti}, p_{tj})}{|\vec{p}_{ti} + \vec{p}_{tj}|} < z_{\text{cut}}, \quad \text{and} \quad \Delta R_{ij} > R_{\text{prune}}. \tag{4.9}$$

If this is the case, the two constituents are not merged into a single jet, but the one with smaller transverse momentum is discarded. This suppresses splittings induced by QCD radiation, where the soft singularity of the splitting functions mainly gives rise to asymmetric splittings. The set of all surviving subjets constitutes the *pruned* jet. The value of R_{prune} is not fixed, but is set dynamically. A common choice is $R_{\text{prune}} = m_{\text{jet}}/p_{\text{t,jet}}$, where m_{jet} and $p_{\text{t,jet}}$ are the jet invariant mass and transverse

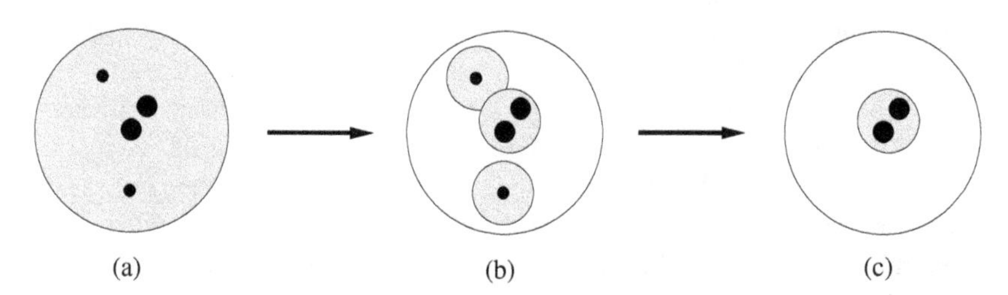

Figure 4.6. (a) A jet of large radius R containing two hard constituents (the big black circles) and two soft constituents (the small black circles). (b) The jet is reclustered with radius $R_{\text{sub}} < R$, and three subjets are found. (c) All subjets such that $p_{tj} < z_{\text{cut}}\Lambda_{\text{hard}}$ are discarded, and only the subjet containing the two hardest constituents survives.

momentum, respectively, *before* checking the conditions in equation (4.9). The behaviour of pruning is pictorially illustrated in figure 4.7 for the same configuration considered for trimming. In the shown example, the outcome of the procedure is again a jet containing only the two hardest constituents.

Mass-drop tagger (MDT). In its original formulation, the mass-drop tagger [20] is aimed at finding a hard jet that originates from the two-body decay of a massive particle, like the Higgs or a vector boson. Hence, it necessarily contains both grooming and tagging features. We will now concentrate on the grooming aspects of MDT, and leave the tagging features for the next section. The procedure starts by clustering an event into fat jets. Then, one reclusters each fat jet using the Cambridge/Aachen algorithm, whose clustering sequence is close to reversing the angular-ordered branching predicted by QCD. Differently from pruning, which modifies the clustering sequence, MDT works backwards. In fact, for each fat jet above a certain p_t threshold, the last step of the clustering is undone, giving two subjets p_1 and p_2, with masses m_{j1} and m_{j2}. In the original version of the MDT, one imposes an asymmetry condition, which is in fact a groomer:

$$\frac{\min\left(p_{t1}^2, p_{t2}^2\right)}{m_{\text{jet}}^2}\Delta R_{12}^2 > y_{\text{cut}}. \tag{4.10}$$

If this is not met, the subjet with the smaller invariant mass is eliminated, and one undoes one more step of the clustering procedure. The condition in equation (4.10), very much like pruning, aims at suppressing asymmetric splittings induced by QCD radiation. Note that, following the study of [22], the grooming procedure described above has been slightly modified with respect to its original version. In the updated version, known as the modified MDT (mMDT), if the condition in equation (4.10) is not met, the jet p_j with the largest transverse mass $m_{tj} = \sqrt{m_j^2 + p_{tj}^2}$ is discarded. This is to avoid following a soft subjet whose mass is generated dynamically by soft-collinear parton branching. The action of the asymmetry condition of the mass-drop tagger is again pictorially explained in figure 4.8. At the end of the procedure, starting the same configuration as in the examples of figures 4.6 and 4.7, one obtains a jet that contains only the two hardest constituents.

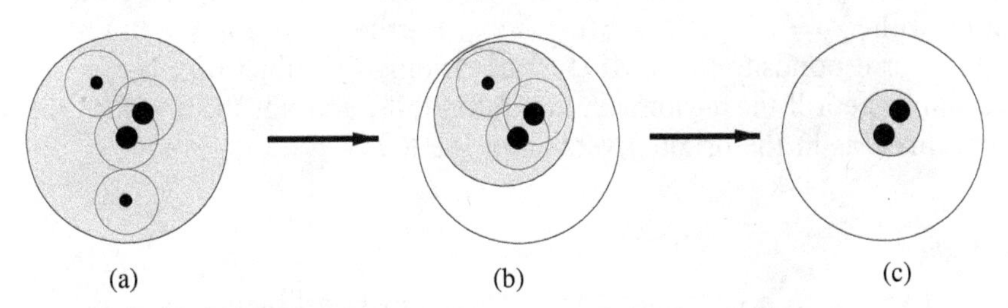

Figure 4.7. (a) The same jet considered in figure 4.6, where circles of radius R_{prune} are drawn around each constituent. (b) First stage of the clustering: the soft jet at the bottom is discarded. (c) Second stage of the clustering: the soft jet at the top is discarded, and one is left with a jet of radius R_{prune} with the two hardest constituents only.

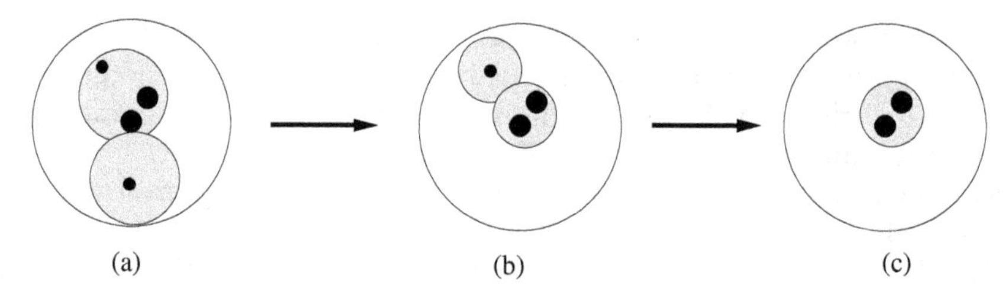

Figure 4.8. (a) The same jet considered in figure 4.6. In the MDT procedure, the last stage of the clustering is undone, and two subjets are found. (b) The asymmetry condition in equation (4.10) is not satisfied, so the soft jet at the bottom is discarded. A further stage of the clustering is undone, and one is left with two subjets. (c) The soft jet at the top is discarded because of the failure of the asymmetry condition, leaving a jet containing only the two hardest constituents.

Soft drop. A further improvement of the modified MDT (mMDT) is the 'soft drop' [21]. Its first step proceeds exactly as the MDT, in that it reclusters a jet with radius R with the Cambridge/Aachen algorithm, and undoes the last step of the clustering of the jet, finding two subjets p_1 and p_2. Then, one checks if they satisfy the soft-drop condition

$$\frac{\min(p_{t1}, p_{t2})}{p_{t1} + p_{t2}} > z_{\text{cut}}\left(\frac{\Delta R_{12}}{R}\right)^{\beta}, \tag{4.11}$$

where β is a real parameter. If the soft-drop condition is not met, then the subjet with lower transverse momentum is removed. A peculiarity of the soft-drop procedure is that it acts either as a groomer or a tagger, according to the value of β. Specifically, for $\beta \geqslant 0$ it removes the soft large-angle constituents of a jet, with β essentially controlling the angular size of the groomed jet. The soft-drop condition for $\beta = 0$ corresponds to the asymmetry condition of the MDT in equation (4.10), which in fact can be rewritten as

$$\frac{\min(p_{t1}^2, p_{t2}^2)}{m_{\text{jet}}^2}\Delta R_{12}^2 \simeq \frac{\min(p_{t1}^2, p_{t2}^2)}{p_{t1}p_{t2}} \simeq \frac{\min(p_{t1}, p_{t2})}{\max(p_{t1}, p_{t2})} > y_{\text{cut}}, \tag{4.12}$$

where we have assumed that the two subjets p_1 and p_2 are very close in angle. In this limit, $\max(p_{t1}, p_{t2}) \simeq \max(z, 1 - z)(p_{t1} + p_{t2})$, where z is the fraction of the jet total energy carried by one of the jets. Therefore, equation (4.12) represents a soft-drop condition with $z_{\text{cut}} = \max(z, 1 - z)y_{\text{cut}} \sim y_{\text{cut}}$. Furthermore, for $\beta = 0$, the soft-drop procedure corresponds to the mMDT, because only the subjet with larger transverse momentum is kept if the asymmetry condition fails, and not the one with the larger invariant mass as in the original version of the MDT.

4.2.2 Taggers

Once hard jets have been cleaned from their soft constituents, one still faces the problem of distinguishing between jets originating from the decay of a heavy object and QCD jets, whose mass is dynamically generated through radiation. The most obvious form of tagging is to keep every jet above a certain transverse momentum threshold that passes the grooming conditions. One can then bin the invariant mass of each jet, hoping to find a peak above the QCD background. The latter should be reduced thanks to the grooming procedure. Taggers actually do much more, as they further probe the properties of each jet to see whether it could originate from the decay of a heavy objects or from QCD. There exists an enormous variety of taggers (see e.g. [13] for a recent review). Here we will just discuss the main ideas behind them, by exploring the features of a limited number of pioneering examples.

Mass-drop tagger. The actual mass-drop tagger is a condition that applies to a pair of subjets of a fat jet that have passed the required of equation (4.10). This already implies that we have two subjets with comparable energy, as expected from the decay of a heavy object. If these are the decay products of a heavy particle, if one eliminates either of the two, the other subjet will be almost massless. Therefore we will observe a drop in the jet invariant mass. If instead the invariant mass does not decrease significantly, it means that the subject with the larger invariant mass can be declustered further to find quasi-massless constituents. In practice, one imposes the mass-drop condition on two subjets with invariant masses m_{j_1}, m_{j_2} as follows

$$\max(m_{j_1}, m_{j_2}) < \mu \, m_{\text{jet}} \tag{4.13}$$

with μ some real positive parameter. If this condition is satisfied, the two subjets are kept and the jets is tagged. If this is not satisfied, the modified MDT requires that the jet with the lowest transverse mass is discarded. Then, a new declustering is performed, two subjets are found, and the conditions in equations (4.10) and (4.13) are checked again. If neither condition is ever satisfied at any stage of the declustering, the jet is discarded as a whole. We can picture the effect of the mass-drop condition in panel (c) of figure 4.8. There, after grooming, we have two hard partons left inside the jet. If we perform a further declustering, both subjets will pass the asymmetry condition in equation (4.10). Then, if one eliminates either of the two, one is left with a single massless constituent. At this stage, the mass-drop condition is met and the procedure stops leaving a tagged jet with two massless constituents.

The mass-drop condition implements a basic property of the decay of a heavy particle into two hard massless objects. Therefore, if one is searching for a two-prong decay, i.e. the heavy particle decays into two partons only, the procedure can end here. For three-prong decays, such as those of the top quark, one can consider the subjet with larger invariant mass, and undo the clustering further until another mass drop is found. The mass-drop procedure can then be generalised to more complicated decays.

Notice that the MDT procedure does not need to be necessarily used in conjunction with the asymmetry condition in equation (4.10), but it could be applied to any groomed jet or in addition to a grooming procedure.

Soft drop as a tagger. For $\beta \geqslant 0$, the soft-drop condition in equation (4.11) acts as a groomer. Therefore, the mass resolution of the resulting jet can be improved by imposing the MDT condition in equation (4.13) at each stage of the declustering. On the contrary, the soft-drop condition for $\beta < 0$ avoids the need of the mass-drop check altogether. In fact, for $\beta < 0$, the soft-drop condition implicitly forces two subjets to have a large-angle separation, and hence selects only jets that contain at least two hard, well-separated subjets, as is typical for two-prong decays. Therefore, for $\beta < 0$ the algorithm behaves as a tagger.

Filtering. Tagged jets can be further modified by applying filtering. Although this technique was introduced for the first time in combination with the MDT, it could be supplemented to any other tagger, and applied to any jet, be it groomed or now. Once the tagged jet has been found, it might be necessary to further remove contamination from UE/PU. In doing this, one needs to retain soft constituents arising from soft radiation from the primary hard partons that have initiated the tagged jet. Losing them would result in spoiling the resolution of the jet mass peak, as discussed in section 4.1. Filtering specifically aims at keeping the soft constituents that originated from soft radiation, while at the same time trying to exclude spurious soft jets from a uniform background. This is achieved by reclustering the tagged jet with a smaller radius R_{filt}, and measuring the distribution in the invariant mass of only n_{filt} subjets as the quantity that is supposedly closer to the mass of the heavy particle that originated the jet. Using similar arguments as in section 4.1, it is possible to show that perturbative QCD radiation causes a loss in the jet mass that follows $\ln(1/R_{\text{filt}})$, hadronisation gives also a loss, proportional to $1/R_{\text{filt}}$, whereas a uniform background gives an enhancement in the jet mass that grows like R_{filt}^2 [23] [1]. It is therefore possible to find an optimal radius that minimises the jet-mass resolution. An example on how to determine this optimal radius using QCD considerations is found in section 4.2.3. Historically, the MDT/filtering technique was introduced for the first time to look for a boosted Higgs boson decaying into a $b\bar{b}$ pair, and produced in association with a vector boson at the LHC with $\sqrt{s} = 14$ TeV. The result of that analysis was that it was possible to find a candidate fat jet that survived the MDT procedure, and whose mass distribution after filtering showed a peak around the Higgs mass, taken then to be 115 GeV in [20].

Event shapes as taggers. The procedures we have seen so far are based on the declustering of a jet and subsequent elimination of jet constituents. A complementary approach is to use event shapes, or jet resolution parameters, using the constituents of a fat jet as inputs. Suppose that one reclusters a fat jet using the k_t algorithm. Then, a single jet will have a one-jet resolution d_{01} that is of order one. If a jet is made up of two energetic subjets, then the two-jet resolution d_{12} will be also of order one. A QCD jet instead will tend to have a small two-jet resolution. Therefore, the distribution in d_{12}/d_{01} will be peaked at lower values for QCD jets than for jets originated from hadronic decays of a heavy particle. One can then separate signal from background by just performing a cut in d_{12}/d_{01}. For two-prong decays, this is similar to the asymmetry condition for the MDT in equation (4.10). Similarly, one can look for three-prong decays by performing a cut on d_{23}/d_{12}, and so on. This is the basis of the tagging procedure encoded in the program Y-splitter [24]. Jet resolutions are just an example of a discriminating variable. More recent studies use the N-subjettiness variable [25, 26], defined after declustering a jet of radius R into N subjets as

[1] Note that, in filtered jets, the enhancement in the jet invariant mass due to UE/PE grows like R_{filt}^2, and not R_{filt}^4 as expected. This improvement is a direct consequence of the filtering procedure.

$$\tau_N^{(\beta)} = \frac{1}{d_0} \sum_{i \in \mathrm{jet}} p_{ti} \min\{(\Delta R_{1i})^\beta, (\Delta R_{2i})^\beta, \ldots, (\Delta R_{Ni})^\beta\}, \quad d_0 = \sum_{i \in \mathrm{jet}} p_{ti} R^\beta, \tag{4.14}$$

with β a real parameter that gives the additional freedom to vary the relative importance of soft and collinear radiation, similar to the homonymous parameter for the soft drop. The sums in the above expression run over all jet constituents, and ΔR_{jk} is the y–ϕ distance between the jet momentum p_j, with $j = 1, \ldots, N$, and the jet constituent p_k. The use of N-subjettiness is similar to that of the Y-splitter: an N-prong decay can be selected via a suitable cut on τ_N/τ_{N-1} (with $\tau_N \equiv \tau_N^{(1)}$). N-subjettiness is more tractable than the Y-splitter from an analytical point of view, because of the existence of all-order factorisation formulae in SCET for its global variant N-jettiness [27].

Another example of variables that can be used to investigate jet substructure is ratios of generalised energy-correlation functions ECF(N, β) [28], defined by the relations

$$\begin{aligned}
\mathrm{ECF}(0, \beta) &= 1 \\
\mathrm{ECF}(1, \beta) &= \sum_{i \in \mathrm{jet}} p_{ti} \\
\mathrm{ECF}(2, \beta) &= \sum_{i<j \in \mathrm{jet}} p_{ti} p_{tj} \left(\Delta R_{ij}\right)^\beta, \\
\mathrm{ECF}(N, \beta) &= \sum_{i_1, i_2, \ldots, i_n \in \mathrm{jet}} \left(\prod_{a=1}^{N} p_{ti_a}\right)\left(\prod_{b=1}^{N-1} \prod_{c=b+1}^{N} \Delta R_{i_b i_c}\right)^\beta .
\end{aligned} \tag{4.15}$$

For instance, to select two-prong decays one can consider the ratio $C_2^{(\beta)}$, defined as

$$C_2^{(\beta)} = \frac{\mathrm{ECF}(3, \beta)\mathrm{ECF}(1, \beta)}{\mathrm{ECF}(2, \beta)^2}. \tag{4.16}$$

By comparing equation (4.15) and (4.14), one sees that $C_2^{(\beta)}$ is closely related to the 2-subjettiness variable defined in equation (4.14). A comparison of the performance of generalised energy-correlation functions and N-subjettiness can be found in [28].

Various generalisations are possible from the energy-correlation functions in equation (4.15). For instance, one can take ratios of ECFs with different powers of β. For instance, a variable used to select two-prong decays is [29]

$$D_2^{(\alpha,\beta)} = \frac{\mathrm{ECF}(3, \alpha)}{\mathrm{ECF}(2, \beta)^{3\alpha/\beta}}. \tag{4.17}$$

One can also construct the so-called 'generalised energy flow polynomials' (EFPs), defined by [30]

$$\mathrm{EFP}_G = \sum_{i_1, i_2, \ldots, i_n} z_{i_1} z_{i_2} \cdots z_{i_n} \prod_{(k,l) \in G} \theta_{i_k i_l}, \tag{4.18}$$

where G is any subset of particle pairs, which can in fact represented as a two-dimensional graph. The sum in equation (4.18) to all particles in a jet, although there is nothing that prevents considering different subsets of particles. The function z_i is a placeholder for the energy fraction of particle p_i, for instance $p_{ti}/p_{\mathrm{t,jet}}$, and θ_{ij} a suitable collinear safe distance between particles p_i and p_j, for instance $\Delta R_{ij}/R$, with $p_{\mathrm{t,jet}}$ and R the jet transverse momentum and radius respectively. All the jet observables considered so far are IRC safe. As such, they are not very sensitive to collinear emissions. One can enhance this sensitivity by considering also collinear unsafe quantities, like the generalised angularities [31]

$$\lambda_\beta^\kappa = \sum_i z_i^\kappa \theta_i^\beta, \tag{4.19}$$

for $\kappa > 1$. Here, z_i is again the energy fraction of particle p_i, and θ_i a collinear safe angular distance from a set axis. All these variables are the basic building blocks for many jet-substructure studies or for strategies to discriminate between quark and gluon jets, as discussed in section 4.3.2.

The onset of Machine Learning. Besides the methods presented here, many more groomers and taggers have been proposed. Given such a plethora of approaches, it might be useful to combine them in various ways. This can be done manually, by combining the methods that seem more suitable for a given experimental analysis, or automatically with Machine Learning (ML). For instance, one could feed all groomers and taggers to a neural network, and let the network find the better discriminant. A pioneering example of this approach is provided by [32]. This use of ML uses high-level input for the neural network. There exist various other methods that input low-level information, for instance particle momenta or even the output of the detectors. One of the most popular, which is widely used to benchmark tagger performances, is ParticleNet [33]. What one wants is to maximise both the signal acceptance and the background rejection rate. These two quantities are plotted one against the other in the so-called receiver operating characteristic (ROC) curve (see section 4.3 on quark–gluon jet discrimination for concrete examples). Although ML is invaluable to optimise the performance of various taggers, it is still the task of the user to decide which observables or strategies should be exploited. We can gain invaluable insight on the performance of a given groomer or tagger if we study the way it probes QCD radiation. This is the topic of the next section. As this discussion is more theoretical in nature, it can be skipped by a reader interested only in the qualitative features of the various taggers.

4.2.3 QCD insights into jet substructure

Given the variety of methods available to investigate jet substructure, it is natural to ask the question on how to assess their performance. One possibility is of course to construct the ROC curve corresponding to a given tagger. However, such trials are based on Monte Carlo event generators, which sometimes do not reproduce exactly the features predicted by perturbative QCD. It is therefore useful to gain insight on the features that taggers should display in QCD. As these procedures are quite involved, analytic calculations are in general technically demanding, and at a lower accuracy with respect to global jet observables. Therefore, it is useful many times to consider simplified models that represent good approximations of the actual taggers. Once some relevant properties have been obtained with analytic techniques, it is important to check that those are reproduced by the Monte Carlo event generators that will be used for actual experimental analyses.

4.2.3.1 Invariant mass of groomed QCD jets

A proposal to assess the performance of different groomers [22] is to calculate the invariant mass distribution of a groomed QCD jet in perturbative QCD. In particular, the authors of [22] considered the distribution in the dimensionless variable $\rho = m_{\text{jet}}^2/(p_{t,\text{jet}}^2 R^2)$, after trimming, pruning or the asymmetry condition of the MDT. In general, grooming will reduce the fraction of QCD jets contributing to the invariant mass distribution. On top of this, one would prefer the mass distribution of a QCD jet to be featureless, without peaks that can mimic a signal peak. For instance, the plain jet-mass distribution, i.e. that of a jet where no grooming has been applied, has a peak at around 10% of the jet transverse momentum. This peak is a consequence of the fact that QCD radiation that dominates the mass distribution at large invariant masses, is suppressed and vanishes at low invariant masses due to the presence of a Sudakov form factor, as explained in section 3.2, thus generating a peak. We call this feature a Sudakov peak. A first interesting result of a QCD analysis of the ρ distribution is that, after trimming or pruning, it still shows features. This can be seen for instance in figure 4.9, which shows the ρ distribution for trimmed (left) and pruned (right) quark jets. In particular, it is interesting to investigate the behaviour of the distribution $\rho d\sigma/d\rho$ for decreasing values of ρ. As expected, the distribution starts with an increase, due to the jet being dominated by a single gluon emission. Then, at around $\rho \simeq 0.1$, the distribution becomes almost flat until a turning point at $\rho \simeq 0.01$, which,

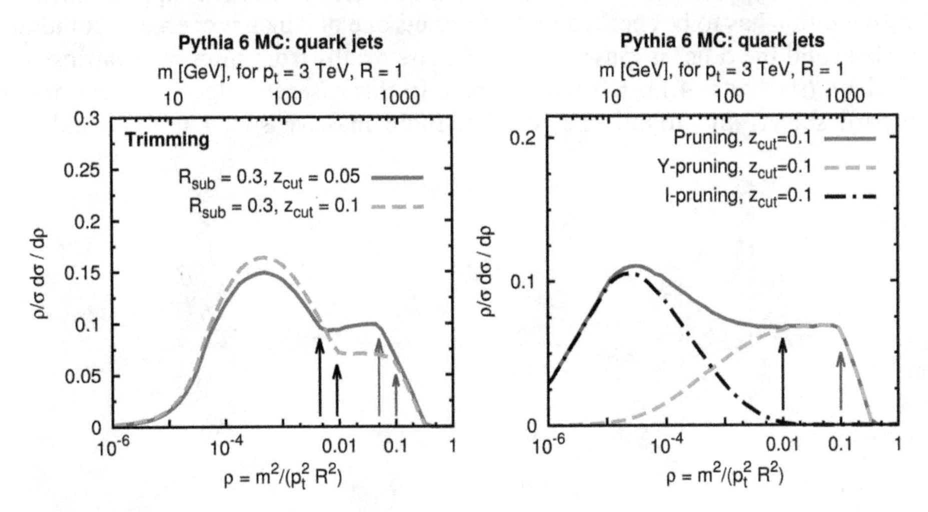

Figure 4.9. The distribution $\rho d\sigma/d\rho$ for a trimmed (left) and a pruned (right) quark jet with $p_t = 3$ TeV [22]. The plots report also the corresponding value of the jet mass, here denoted by m. Reproduced from [22] (2013). With permission of Springer.

for 1 TeV jets is of the order of the electroweak scale, where the masses of interesting particles lie. The plots in the figure are obtained via an analytic calculation using perturbative QCD, but the same features are correctly reproduced by popular Monte Carlo event generators. This feature can be easily understood for a trimmed jet, which differs from the original jet just by the fact that it has a smaller radius. This gives a peak that is shifted to lower values of ρ with respect to that of the plain jet-mass distribution. In the case of pruning the situation is more complicated, and requires a detailed calculation at order α_s^2. In fact it turns out that pruning actually selects two kind of jets, called I-pruned and Y-pruned jets, as suggested by their shape. If one does not find in any of the clusterings a pair of subjets p_i and p_j such that $\Delta R_{ij} > R_{\text{prune}}$ and $\min(p_{ti}, p_{tj}) > z_{\text{cut}}|\vec{p}_{ti} + \vec{p}_{tj}|$, it means that the jet mass m_{jet} that determines $R_{\text{prune}} = m_{\text{jet}}/p_{\text{t,jet}}$ is set by the energy of the last soft large-angle constituent that is pruned away (see figure 4.10, left-hand panel). Therefore, the mass of the pruned jet will be just the mass of a jet of radius R_{prune}, which has a peak at some low value of ρ, which is what is seen in figure 4.9. These narrow jets that are called I-pruned jets, due to their pencil-like shape. If, however, one finds at least one clustering with $\Delta R_{ij} > R_{\text{prune}}$ and $\min(p_{ti}, p_{tj}) > z_{\text{cut}}|\vec{p}_{ti} + \vec{p}_{tj}|$, the jet will have a two-prong structure, and its invariant mass, and hence R_{prune}, will be basically unaffected by additional soft radiation (see figure 4.10, right-hand panel). Such jets are called Y-pruned jets, again due to their shape. I-pruned jets can be discarded by looking at the clustering sequence, leaving only Y-pruned jets. The mass distribution of Y-pruned jets no longer has a Sudakov peak, but vanishes smoothly with ρ. Interestingly, although pruning was designed as a groomer, only I-pruning is a groomer, whereas Y-pruning is in fact a tagger, since only two-prong structures are kept. Concerning MDT, the analysis of [22] discovered a flaw in the original procedure, which has now been replaced by the mMDT described in section 4.2.1. This problem appears again for the first time at order α_s^2. In fact, suppose the mass-drop condition has to be checked on two subjets one of which is energetic but almost massless, and the other is massive, but its mass results from multiple splittings of a soft gluon (see figure 4.11, left-hand panel). In this case, the algorithm will discard the hard subjet, and follow the soft but more massive subjet until it finds two

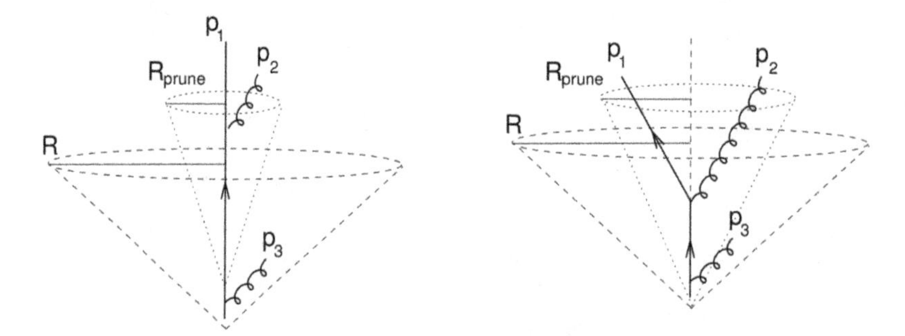

Figure 4.10. Configurations at $\mathcal{O}(\alpha_s^2)$ that give rise to a I-pruning (left) and Y-pruning (right). Left: soft gluon p_3 sets R_{prune}, and is then pruned away. Right: R_{prune} is set by hard gluon p_2, and soft gluon p_3 is just pruned away. Both pictures have been drawn by the author, following the pictures in [22].

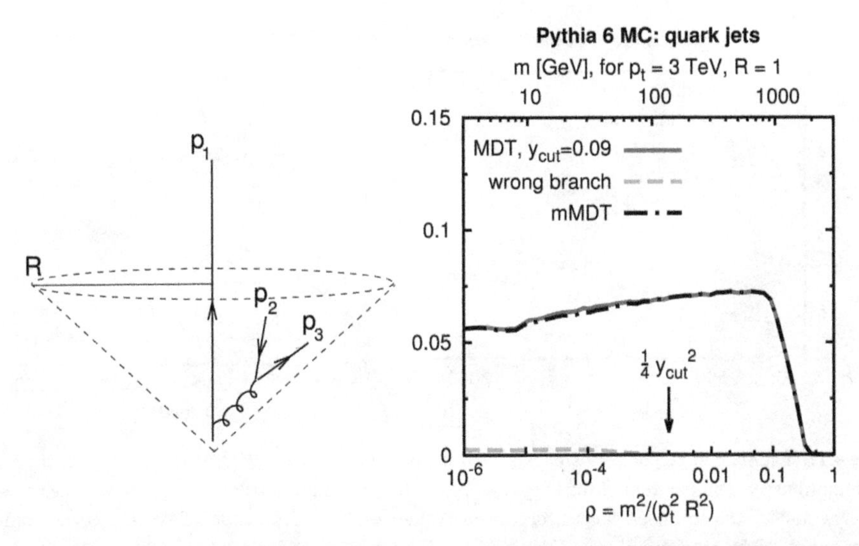

Figure 4.11. Left: a configuration that leads to the original mass-drop procedure to follow a soft branch (labelled 'wrong' in the right-hand panel); in this case the original algorithm discards the subjet made up of a single energetic parton, because it is energetic, but massless. The picture has been drawn by the author, and is similar to the corresponding picture in [22]. Right: the distribution $\rho d\sigma/d\rho$ for a jet tagged with the MDT procedure. Reproduced from [22] (2013). With permission of Springer.

massless constituents. This occurrence is rare, because soft subjets are normally eliminated by the asymmetry condition in equation (4.10). However, this is clearly an unwanted feature of the MDT, and can be avoided by just discarding the subjet with the lower transverse mass, instead of that with the lower invariant mass. Another possibility is to follow the branch with the larger transverse momentum, as is done by the soft-drop procedure. One more interesting feature of the mMDT is that it is possible to tune its parameter y_{cut} in such a way that the mass distribution of a tagged jet does not have the characteristic Sudakov peak, being more or less flat from the point in which the MDT condition sets in (see figure 4.11, right-hand panel). Such featureless mass distribution is ideal to distinguish signal from background. From this analysis one sees that an analytical understanding of groomers and taggers from the point of view of perturbative QCD can lead to improvements that would be difficult to devise through a trial-error procedure. Similarly, one can investigate the behaviour of the soft-drop tagger as a function of β. As already stated, for $\beta > 0$ the procedure is basically equivalent to trimming or pruning, for $\beta = 0$ it roughly corresponds to the mMDT, whereas for $\beta < 0$, the extreme tagging mode, its behaviour resembles that of Y-pruning.

Such QCD analyses make it possible to understand the performance of different taggers. In figure 4.12 we show a couple of examples of such curves for the tagging of a boosted W boson decaying hadronically. The main quantities considered in both analyses are the acceptance efficiency for the signal ε_{S} (the number of tagged signal jets over the total number of signal jets) versus that for the background ε_{B}. The curve on the left-hand panel of figure 4.12, taken from [22], shows $\varepsilon_{\mathrm{S}}/\sqrt{\varepsilon_{\mathrm{B}}}$ as a function of the tagged jet transverse momentum. There one can see the better performance of

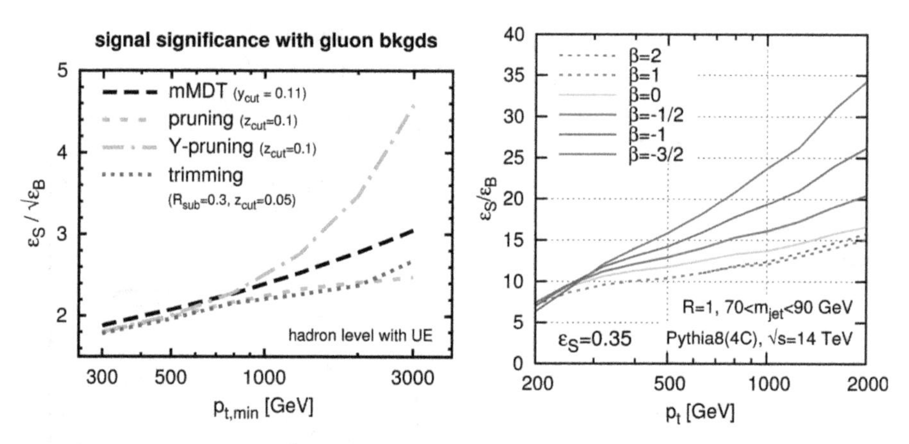

Figure 4.12. Estimate of signal versus background tagging efficiency for various tagger as a function of the minimum transverse momentum of the tagged jet $p_{t,\min}$ (left, reproduced from [22] (2013), with permission of Springer), and of the soft-drop tagger for various values of β, as a function of the tagged jet transverse momentum p_t (right, reproduced from [21] (2014), with permission of Springer).

taggers with respect to pure groomers, with Y-pruning outperforming the other taggers at large p_t. The plot on the right-hand panel, taken from [21], shows instead $\varepsilon_S/\varepsilon_B$ as a function of the jet transverse momentum, for different variants of the soft-drop procedure, corresponding to different values of the parameter β. The pattern is very similar, with taggers performing better than pure groomers. Also, variants with negative β outperform the one with $\beta = 0$, which corresponds roughly to the mMDT. Their performance is similar to that of Y-pruning. This is not surprising, in that setting $\beta < 0$ *de facto* removes jets that are very collimated, thus leaving two-prong structures only, the more aggressive the cut the more negative β is.

Last but not least, it is now possible to compute the invariant mass of groomed QCD jets at very high accuracy in perturbative QCD [34–36]. These achievements, together with the extraordinary precision reached by LHC experiments, makes it possible to open new avenues for probing QCD dynamics in a regime never explored before. This will be discussed in more detail in section 4.2.4, which describes the current status of jet-substructure studies at the LHC.

4.2.3.2 Colourless particles and filtering

In section 4.2.1 we have presented the filtering procedure, that aims at further cleaning an already groomed and tagged jets. The choice of the number of subjets to be kept while filtering, as well as the radius R_{filt} of the filtered subjets could be informed by QCD consideration. Here we consider the pioneering example of [23], which informed the choices of [20]. We then focus on the hadronic decay of a colourless particle, in particular a Higgs boson decaying into a $b\bar{b}$ pair. Since the Higgs boson is a so-called 'colour singlet', i.e. it has no colour charge, QCD coherence offers an important handle to determine the optimal filtering radius. In fact, suppose a tagger has been able to find two subjets separated by a distance R_{jj}. Then, the coherence properties of QCD radiation will force an extra soft gluon to be radiated within two cones of radius R_{jj}, centred in each of the two hard subjets. Therefore, R_{filt} will have to be a fraction of R_{jj}. Furthermore, it is possible to perform a number of analytical and semi-numerical calculations to determine the dependence of the jet-mass resolution on R_{filt} [23]. In the specific case a of Higgs boson decaying into a $b\bar{b}$ pair, with $n_{\mathrm{filt}} = 3$ one can observe a minimum in the jet-mass resolution as a function of $\eta = R_{\mathrm{filt}}/R_{jj}$, located at $\eta = 0.3$. This motivates the value $R_{\mathrm{filt}} = \min\{0.3, R_{b\bar{b}}/2\}$ and $n_{\mathrm{filt}} = 3$ proposed in [20].

4.2.3.3 Interjet radiation and the pull

Let us consider again the case of a boosted Higgs decaying into a $b\bar{b}$ pair and consider a tagged jet, with two subjets, both tagged as b-jets (pictorially represented in the left-hand panel of figure 4.13). An irreducible background to this signal originates a gluon splitting into a $b\bar{b}$ pair. Due to the collinear singularity of gluon splitting, it is presumable that such background will be reduced by the grooming procedure, for instance imposing the asymmetry condition in equation (4.10). However, QCD jet production has a larger cross section than electroweak processes, so many events with two well-separated b-jets arising from gluon splitting are likely survive the grooming procedure. A way to reduce this background is to observe that a gluon carries a colour charge, whereas a Higgs does not. Therefore, due to the coherence properties of QCD radiation, in the case of $H \to b\bar{b}$, radiation coming from the $b\bar{b}$ pair will be typically contained inside the fat jet (see again figure 4.13), whereas in the gluon case there will be a considerable amount of radiation outside as well. A way of exploiting this information is to veto additional jets in the event. In fact, final-state radiation at angles larger than the jet radius will produce additional jets, so that vetoing this activity suppresses the background through a Sudakov form factor, whereas no such price is paid for the signal. However, both signal and background events are affected by initial-state radiation, so that Sudakov form factors will suppress the signal as well. Imposing vetoes on extra jets is a way to exploit the angular distribution of energy-momentum flow in QCD and Higgs jets. In fact, in $H \to b\bar{b}$, QCD radiation will be contained in the region between the two b-jets (we say in this case that the b and anti-b are 'colour connected'), whereas in a

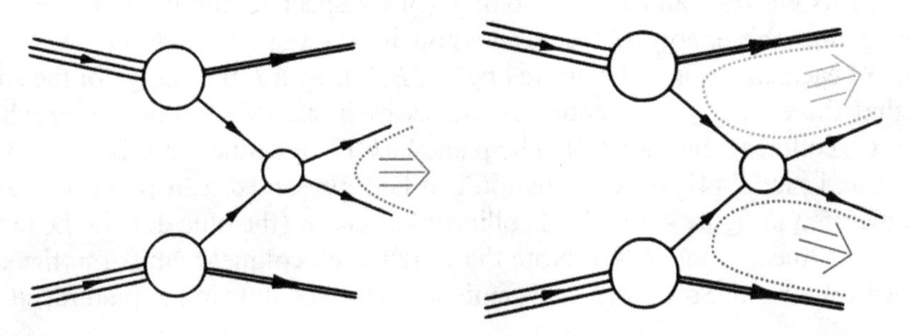

Figure 4.13. A pictorial representation of the preferred directions of QCD radiation (the big arrows) in an event in which a $b - \bar{b}$ pair is produced via the decay of a colourless particle (left), or through the splitting of a coloured particle, for instance a gluon (right). Reprinted (figure 1) with permission from [37], Copyright (2010) by the American Physical Society.

case of a QCD event, radiation will preferably occupy the region between each jet and the closest beam. This is pictorially illustrated in figure 4.13. This effect can be quantified through the so-called 'pull' vector. This is defined for each jet, and aims at identifying the direction in which colour is 'pulled'. For any particle p_i in a jet J of momentum p_J, one considers the two-dimensional distance vector $\vec{r}_i = (y_i - y_J, \phi_i - \phi_J)$. In terms of this quantity the pull vector of jet J is defined as

$$\vec{v}_p^J \equiv \sum_{i \in J} \frac{p_{ti} |\vec{r}_i|}{p_{tJ}} \vec{r}_i. \tag{4.20}$$

An interesting quantity to look at is the pull angle θ_P that the pull vector forms with respect to a reference axis. In [37] the reference axis was chosen to be that of one of the beams. The expectation was that the distribution in the pull angle of a b-jet in signal and background events would be peaked at $\theta_t = 0, \pm\pi$ for background events, and at the position of the other b-jet for signal events. This is to a good extent confirmed by data, as will be discussed in section 4.2.4. Note that the pull angle is not IRC safe, as it does not vanish in the presence of an infinitely soft emission [38]. It is however possible to construct IRC safe projections of the pull vector that still maintain good sensitivity to colour flow [38].

4.2.3.4 The Lund jet plane

As we have seen multiple times in different contexts, if we plot the constituents of a fat jet in the y–ϕ plane, and assign colours to pixels in that plane, we obtain an image. We can then use convolutional neural networks to distinguish 'images' arising from QCD jet from those due to the hadronic decay of a boosted heavy particle. This is the approach pioneered in [39, 40].

The question arises whether this representation is appropriate for QCD jets, which are mostly made up by a small number of energetic partons accompanied by many softer ones. As discussed in chapter 3, soft radiation is distributed logarithmically in the energy and angle with respect to its hard emitter. We can conveniently represent emissions close to a given hard emitting parton (leg) as points in the plane $\ln(k_t) - \eta$, where k_t is the relative transverse momentum of the emitted particle with respect to its emitter, and η the rapidity with respect to the emitter's direction[2]. Imposing that the energy of each emission is less than its emitter, leads to the rapidity of each emission to be limited by $\ln(2E/k_t)$, with E the energy of the emitter. Note that the energy of the emitter changes by a sizeable fraction only when an emission is collinear, but not soft. The plane $\ln(k_t) - \eta$ defined in this way is known as the 'Lund plane' [41]. Each emission can be represented as a point in the Lund plane, pictured in figure 4.14. Hard collinear emissions (the blue dots in the picture), sit along the line $\eta = \ln(2E/k_t)$. Note that, after each collinear emission, the energy of the leg is reduced. Soft large-angle emissions (the red dots in the picture) sit along

[2] The definition of an emitter's direction is not unique, and corresponds to fixing a so-called 'recoil scheme'. The relevant condition to define an emitter's direction is that the matrix element for an emission becomes singular when the transverse momentum with respect to its emitter vanishes.

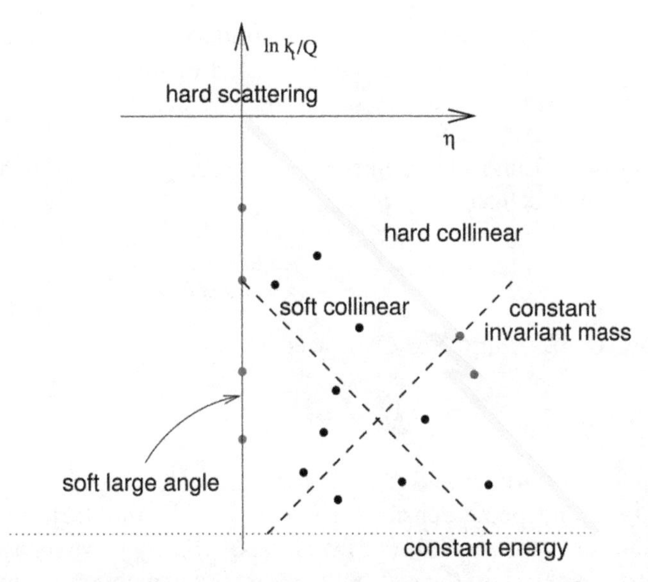

Figure 4.14. The Lund plane for emissions off a single leg.

the line $\eta = 0$. Soft-collinear emissions are the black dots in between the hard collinear and the soft large-angle boundaries. Lines parallel to the collinear boundary represent contours with constant energy, whereas lines orthogonal to those are contours of constant invariant mass. Also, QCD dynamics dictates that soft emissions are generated uniformly in the Lund plane. Therefore, the Lund plane is extremely useful theoretically for resummed calculations, as areas represent double logarithms, lines single logarithms, and dots contributions of relative order α_s. One can also represent secondary emissions, i.e. those that do not arise from a hard parton directly produced in a collision, by introducing secondary Lund planes.

If we consider any jet, we expect its constituent subjets to be closely related to the emissions that have originated them. Therefore, it is useful to find an explicit correspondence between emissions and subjets. One proposal is to construct the Lund jet plane, whose points represents subjets of a given fat jet [42]. The starting point is the same as for other jet-substructure procedures. One considers a fat jet and reclusters its constituents with the Cambridge/Aachen algorithm. Then, the last clustering is undone, giving two subjets, with momenta p_a and p_b and $p_{ta} > p_{tb}$. Then, a number of variables are associated to this declustering:

$$\begin{aligned} &\Delta \equiv \Delta R_{ab}, \quad k_t \equiv p_{tb}\Delta, \quad m^2 \equiv (p_a + p_b)^2, \\ &z \equiv \frac{p_{tb}}{(p_{ta} + p_{tb})}, \quad \kappa \equiv z\Delta, \quad \psi \equiv \arctan\frac{y_b - y_a}{\phi_b - \phi_a}. \end{aligned} \tag{4.21}$$

If p_b is a single parton, k_t and $\ln(1/\Delta)$ are the Lund variables associated to that parton. Therefore, we can start mapping the jet p_b in the Lund jet plane $\ln(k_t) - \ln(1/\Delta)$. The procedure is then repeated by declustering p_a, the subjet with the largest transverse momentum. This gives access to the primary Lund jet plane. Following p_b would give access to a secondary Lund jet plane. We call each point in the primary Lund jet plane an 'emission'. Emissions defined in this way are by construction IRC safe, as they are subjets defined through an IRC safe clustering procedure.

Given the primary Lund plane, one can construct the average density of points, that for each event is defined as

$$\rho(\Delta, k_t) = \frac{1}{N_{\text{jet}}} \frac{dn_{\text{emissions}}}{d \ln(k_t) d \ln(1/\Delta)}. \tag{4.22}$$

In the soft and collinear limit, for QCD jets we have

$$\rho(\Delta, k_t) \simeq 2C \frac{\alpha_s(k_t)}{\pi}, \tag{4.23}$$

where $C = C_F$ for a quark jet and $C = C_A$ for a gluon jet. Figure 4.15 brings side to side the average Lund plane density for selected QCD and W jets [42], simulated with parton-shower event generators. If we look at QCD jets we recognise a number of expected features. First, in the soft-collinear region, we observe a uniform density, as predicted by QCD. This grows at lower transverse momenta due to the running QCD coupling becoming large. The lower left corner represents the contribution of initial-state radiation, as well as PU/UE. The pattern of radiation in W jets is completely different. There we observe a peak in the collinear region, corresponding the decay of a W boson, with a position determined by the mass of the decaying W boson. We notice also two dips in the density (the whitish regions above and below the mass peak). These reflect the fact that the W boson is a colour singlet, hence QCD radiation too far from its decay products is suppressed due to coherence effects. These basics observations highlight how the Lund-plane density encodes features that are probed by the taggers we have presented so far, but in a more differential way. For instance, from figure 4.15 we can appreciate the effect of the mMDT.

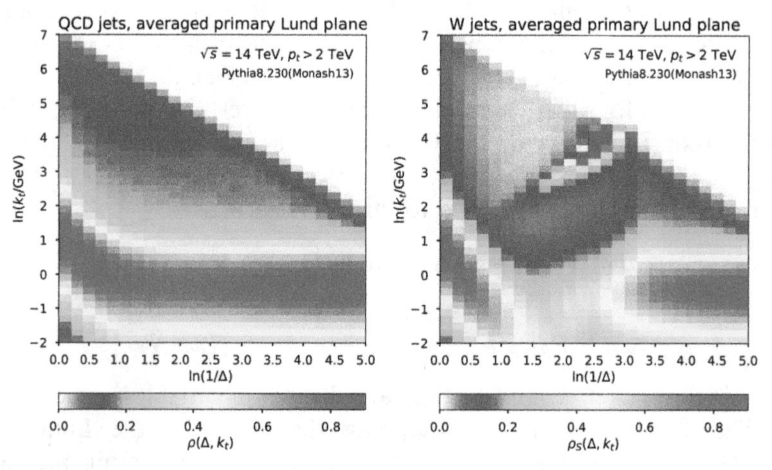

Figure 4.15. The average Lund jet plane density for QCD (left) and W (right) jets. © [42] (2018). With permission of Springer.

This would simply eliminate all emissions below a certain energy, i.e. to the left of a line parallel to the collinear limit. This will leave untouched the peak in the collinear region arising from W decay. More generally, one can use the full information available in the Lund plane to construct both likelihood discriminant, as well as machine learning procedures that use Lund-plane variables as input. The main result is that augmenting information from the Lund plane with machine learning performs better in discriminating between QCD and W jets than using both jet images and event-shape variables as inputs [42].

4.2.3.5 Quantum jets and volatility

In section 2.3 we have presented quantum jets (or Q-jets) as an alternative way of reconstructing jets. In these algorithm, instead of clustering particles according to a certain procedure, which for each event gives rise to a single clustering history, we assign to each recombination a probability. In this way, the same event will have infinitely many clustering histories, each with its own probability. This probability is determined by the rigidity parameter α (see equation (2.15)), and the limit $\alpha \to \infty$ corresponds to the classical case. A consequence of the existence of multiple interpretations in the context of boosted object searches is that signal events that look background-like can be clustered in different ways, and then they have the chance of being kept rather than rejected, as might happen by enforcing a single interpretation. This has the overall effect of increasing the significance of the signal over the background. For instance, figure 4.16 shows the distribution in the invariant mass of a pruned jet, obtained by running several times the classical and quantum versions of the k_t and Cambridge/Aachen jet algorithms. One can see that the classical versions of the algorithms give the same values of the jet mass for each of the considered trees. Note that the two values are different, highlighting the strong dependence on the algorithm of this observable. In the quantum case, for $\alpha = 1.0$, closer to the classical limit, one obtains two distributions, peaked around the corresponding classical values. In a deeper quantum regime, for $\alpha = 0.01$, the difference between the two-jet algorithms is somewhat washed out, and one obtains two very similar distributions. It seems then that quantum jets capture some intrinsic property of these events.

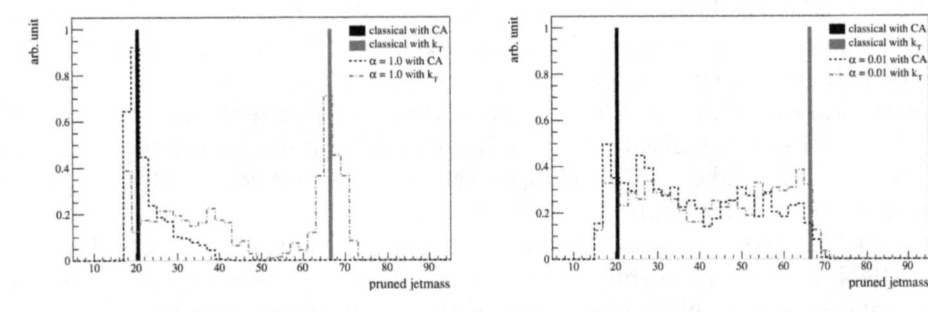

Figure 4.16. The distribution in the invariant mass of a pruned jet for the classical k_t and Cambridge/Aachen algorithms and their quantum versions with the parameter $\alpha = 1.0$ (left) and $\alpha = 0.01$ (right). Reprinted (figure 1) with permission from [43], Copyright (2012) by the American Physical Society.

Quantum jets offer also the possibility of defining new observables, that would be just zero at classical level. For instance, given the mass of a jet m_{jet}, one can define the volatility of a jet as [43]

$$\mathcal{V} \equiv \frac{\sqrt{\langle m_{\text{jet}}^2 \rangle - \langle m_{\text{jet}} \rangle^2}}{\langle m_{\text{jet}} \rangle}, \tag{4.24}$$

where the average has to be taken over all possible interpretations of an event. For instance, if we wish to distinguish between W jets and QCD jets, we expect the former to have a smaller volatility. This is because these jets will have a mass close to the W mass, which will fluctuate less over multiple interpretations. Therefore, performing a cut on volatility is a viable way to discriminate between signal and background [44]. Volatility is currently used among many other taggers in experimental analyses involving boosted objects. An overview of such experimental studies, some of which include volatility, can be found in [45].

4.2.4 Jet substructure studies at the LHC

Ultimately, the effectiveness of all the procedures described so far has to be validated using real data, before it can be used in actual searches. This can be done by considering events in which the decaying heavy particle is known (a so-called 'pure sample'), and performing jet-substructure studies on that sample.

The first two runs of the LHC have been an excellent playground to test the performance of different jet-substructure techniques. A large number of Z and W bosons, as well as top quarks, have offered the possibility to have reasonably pure samples of heavy particles generating two- and three-prong hadronic decays. And of course, a large amount of jet data are the ideal laboratory to investigate the properties of pure QCD jets. Here we will focus on experimental studies of two-prong decays, namely those of W and Z bosons. A recent experimental review on jet substructure, including boosted top decays, can be found in [45]. As usual, we will consider selected examples, which highlight what features of boosted events can be probed currently at the LHC. We will also concentrate on higher energy data with $\sqrt{s} = 13$ TeV.

The ATLAS collaboration extracted a pure sample of W bosons originating from semi-leptonic top-antitop events [46]. This means that the W from one of the tops decays into leptons, i.e. giving a charged lepton (electron or muon) and a neutrino, whereas the W from the other top decays hadronically. This study is able to measure the signal efficiency obtained with a cut based mainly on the variable $D_2^{(\beta,\beta)}$ defined in equation (4.17) (labelled D_2 in [46]), as well as with a deep neural network (DNN) taking as input many shape variables, including the ones we have described in section 4.2. The performance of the two taggers is quite similar (see figure 4.17), with the DNN performing slightly better in the region of intermediate transverse momenta. Given the black-box nature of the DNN, interpreting such a result is not straightforward. However, we remark that a high-level neural network constructs a discriminating variable where observables with the best discriminating

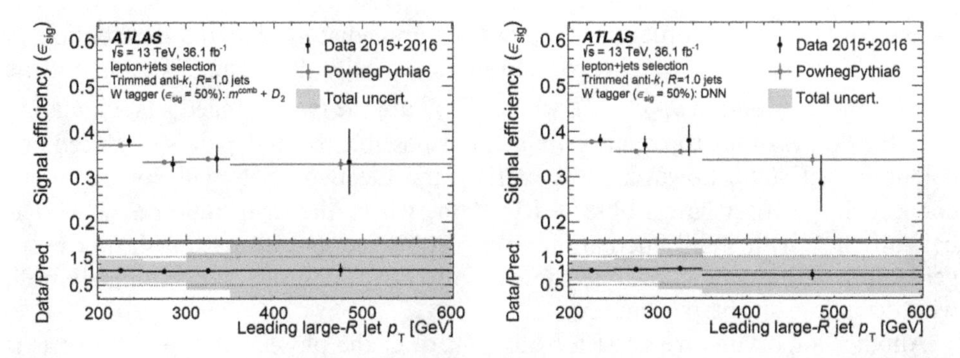

Figure 4.17. The efficiency for detection of W bosons with a cut based analysis based on two variables (left), and on a deep neural network (DNN) (right). © [46] (2019). With permission of Springer.

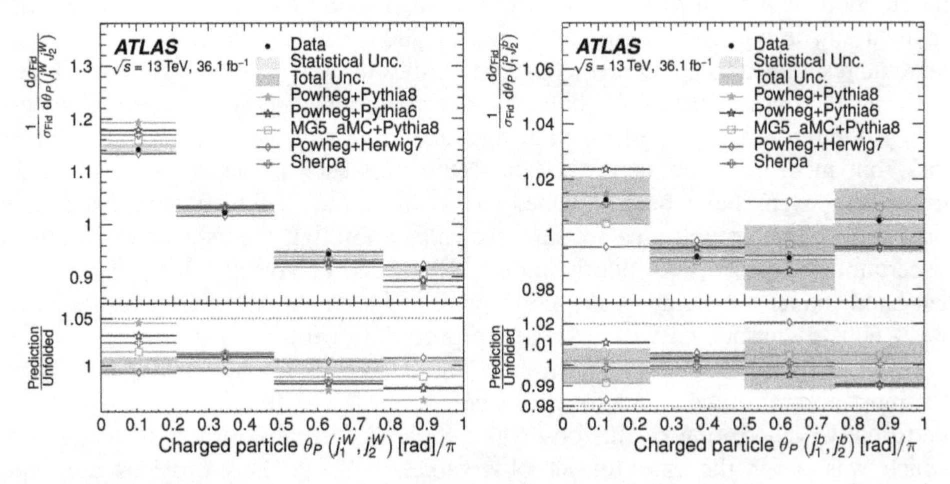

Figure 4.18. The distribution in the angle between the pull vector and the connection vector for two jets originating from the decay of a W boson (left), and two b-tagged jets originating from top decays (right). © [47] (2018). With permission of Springer.

power have the largest weight. In this sense, physics intuition supported suggests D_2 as an optimal variable, and the neural network acts simply as a refinement.

Another important use of $t\bar{t}$ events for jet substructure is to probe colour connections between the jets. This can be done for instance using the pull vector defined in equation (4.20). Reference [47] considers again events corresponding to semi-leptonic $t\bar{t}$ production, and presents measurements of the differential distribution in the angle $\theta_{\mathrm{P}}(j_1, j_2)$ between the pull vector of one jet and a 'connection' vector, joining the two jets in the rapidity–azimuth plane, as well as in the magnitude of the pull vector. For jets that are colour connected, for instance those arising from the decay of a W boson, it is expected that the pull angle will be aligned with the connection vector. This is indeed what is observed in the left-hand panel of figure 4.18, which shows the distribution in $\theta_{\mathrm{P}}(j_1^W, j_2^W)$, where j_1^W and j_2^W are the jets originating from the hadronic decay of a W boson, labelled according to their transverse momentum ordering. One notices immediately that the distribution in $\theta_P(j_1^W, j_2^W)$ is peaked around zero, and monotonically decreasing. This is not the case for the distribution $\theta_P(j_1^b, j_2^b)$, where j_1^b, j_2^b are the two b-tagged jets originating from the decays of the top quarks. In fact, it is possible that a $t\bar{t}$ pair is produced in a colour singlet state, however most of the time the two b-jets will not be colour connected. This is what is observed in data, where the distribution $\theta_P(j_1^b, j_2^b)$ is basically flat, with small fluctuations. We also notice that all parton-shower event generators whose predictions are shown figure (4.18) describe data extremely well, with deviations at the percent level.

Another important stress test for our control of the physics of boosted objects is measurements of distributions in groomed-jet observables. An example of such studies is presented in [48]. There, we find measurements of the invariant mass of a QCD jet after a soft-drop procedure, with different values of the parameters β and $z_{\rm cut}$. The distribution in $\rho \equiv \ln(m^2/p_T^2)$, with m as usual the jet mass and p_T its transverse momentum, is not only compared to parton-shower event generators, but also to analytic resummed to QCD predictions, at various level of accuracy. We remark that such calculations are incredibly challenging from a technical point of view. In fact, not only is the resummation performed at high accuracy, which is in itself a demanding task, but in order to compare to data, NP effects such as hadronisation and the underlying event have been included, as well as the matching with fixed-order predictions. This complicates not only the calculation, but the assessment of theory uncertainties, especially for poorly known NP effects. The result can be found in the left-hand panel of figure 4.19, corresponding to a highly energetic jets with $p_T > 600$GeV, groomed with a soft-drop procedure with $z_{\rm cut} = 0.1$ and $\beta = 0$ (i.e. mMDT). This selection aims at corresponding to the region $\exp(\rho) \ll z_{\rm cut} \ll 1$, where the most accurate predictions for ρ can be obtained. The first observation is that a perturbative description of this observable is valid for many orders of magnitude, which was never the case for jet observables at LEP. This provides a unique opportunity to test our understanding of parton branching and perturbative QCD dynamics in general. On the comparison, we note that NLO is needed to describe the

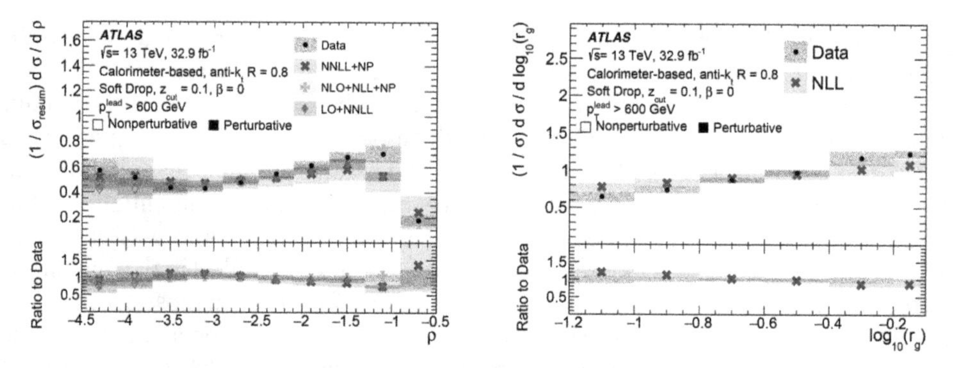

Figure 4.19. The distribution in two-jet observables obtained with a sample of QCD jets groomed with the soft-drop procedure [48]. Details of the jet selection and of the soft-drop parameters can be found in the figure. Reprinted (figures 9 and 10) with permission from [48], Copyright (2020) by the American Physical Society.

region of large ρ. Matching to NLO is provided only by the NLL predictions of [35], although no conceptual issue prevents to match NLO to NNLL resummations. Also, despite the name, the predictions labelled NNLL+NP [36] are in fact NLL accurate. We note that, since the matching is performed at LO only, they fail to describe the large-ρ tail of the distribution. They also have theoretical uncertainty of the order of those of [35], as expected. We note that all presented predictions describe remarkably well the ρ distribution at intermediate and small values, with the NNLL+LO predictions of [34] having the smallest theoretical uncertainties. The right-hand panel of figure 4.19 shows the distribution in r_g, the distance in the rapidity–azimuth plane of the two subjets that survive the soft-drop procedure. The measurements are compared to the NLL resummation of [49], with good, although not perfect, agreement. This calls for higher accuracy of the corresponding theoretical predictions. Distributions in the groomed jet mass have also been measured by CMS [50] over a wide range of jet transverse momenta, showing again remarkable agreement both with parton-shower event generators and analytic calculations.

Another remarkable experimental result is the measurement of the Lund jet-plane density performed by ATLAS [51]. The two variables that are considered are $\ln(R/\Delta R)$, where R is the jet radius, and $\ln(1/z)$, with z the energy fraction of the softer subjet in the C/A declustering procedure defining the Lund plane. The accuracy of the measurements is such that the Lund jet-plane density in slices of the Lund plane can be compared to parton-shower event generators (see figure 4.20), and some of them disagree with data well beyond the experimental uncertainty. This gives plenty of scope for validation of such theoretical tools.

We conclude this section by presenting an example of how boosted object techniques are currently used in new physics searches. This is an analysis performed by CMS [52], looking for a heavy particle W_{KK} decaying into three W bosons. This decay occurs via an intermediate decay of W_{KK} into another heavy particle, the

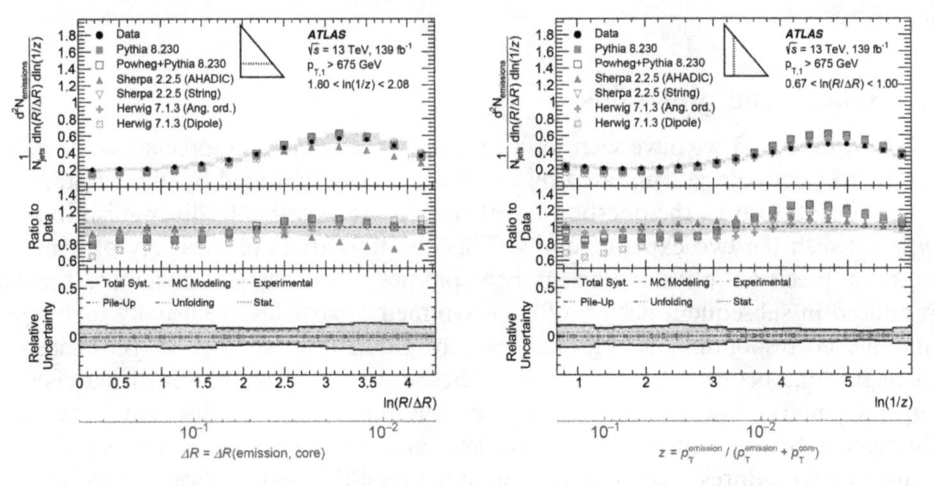

Figure 4.20. The Lund jet-plane density in two slices of the $\ln(R/\Delta R) - \ln(1/z)$ Lund plane, as measured in [51]. The parameters employed for the analysis can be found in the figure. Reprinted (figure 3) with permission from [51], Copyright (2020) by the American Physical Society.

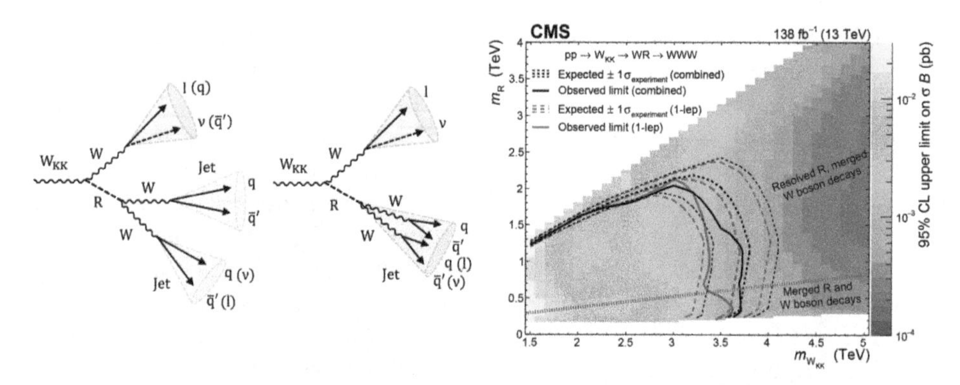

Figure 4.21. Pictorial representation of the decay of a heavy W_{KK} particle (left), and exclusion contours for W_{KK} and a radion. Reprinted (figures 1 and 4) with permission from [52], Copyright (2022) by the American Physical Society.

'radion' R (not to be confused with the jet radius). The analysis considers one W decaying into leptons, while the other two W bosons can emerge either as two boosted jets, or even merged in a single fat jet (see figure 4.21, left-hand panel). For the tagging of boosted W bosons, the machine learning tool DEEPAK8 is employed [53]. This is a low-level tool that take particle momenta (rather their deposit in the detectors) as input. This procedure was previously compared to a variety of taggers based on jet observables. DEEPAK8 was found to outperform observable-based taggers, even when those were used as input to neural networks. By measuring distributions in the reconstructed mass of the decay products of W_{KK} and R, and comparing them with the expectations of various models, it is possible to exclude various regions of masses for W_{KK} and R, as shown in the right-hand panel of figure 4.21. There one can see also the different mass regions that would give rise to two W jets (resolved R) or the two W bosons from the decay of R merged in a single jet (merged R).

4.3 Quark and gluon jets

In many examples we have seen so far, we have been able to appreciate some of the differences between jets initiated by quarks and those initiated by gluons predicted by QCD. This raises the question whether it is also possible to distinguish in some way between the two experimentally. This question arises in a variety of contexts, both in precision physics and in new physics searches. For instance, the jets produced in subsequent decays of supersymmetric particles are usually quark jets, and the corresponding background largely gluon jets (see e.g. [54]). Therefore, distinguishing between the two would increase the discovery power of searches for supersymmetry. This section, the last of this book, aims at discussing the main differences between quark and gluon jets, and how these could be exploited to construct procedures able to discriminate between the two. Although this is at the moment still an open question, it gave rise to inspiring ideas, which constitute somehow an open ending for this book.

4.3.1 Differences between quark and gluon jets

First of all, the flavour of a jet is unambiguously defined only at very high energies, where QCD is free. At any finite energy, any jet originated from a hard parton will also contain other partons radiated from it, as well as from neighbouring partons, hence the flavour of this jet will not be defined in a unique way. A natural definition of the flavour of a jet is the net sum of the flavour of all its constituents. In practice this is an array of dimension equal to the number of quark flavours, where each entry contains the sum of the quarks of a given flavour minus the sum of the corresponding antiquarks, with gluons not contributing to the flavour of a jet at all [55]. If jets are clustered using any IRC safe distance measure, such definition of flavour is invariant under collinear splittings. In fact, a collinear quark and antiquark pair originating from the splitting of a gluon will be clustered in the same jet, thus adding zero to the jet flavour. However, soft large-angle quark–antiquark pairs might be separated by the clustering procedure, and the resulting flavour of jets becomes infrared unsafe. In the case of the k_t algorithm, such unsafety can be removed via a modification of the k_t distance when the softer of the two particles that are to be clustered is a quark. This gives rise to the family of flavour-k_t algorithms, which could have various applications in perturbative QCD studies [55, 56]. Such approach, although theoretically rigorous, cannot be used in practice to determine the flavour of a jet, simply because one does not in general have access to the full information on the flavour of all particles in an event. The second problem is that the algorithm is tailored to the k_t algorithm, while LHC experiments use mainly anti-k_t. So far there exists no established IRC safe algorithm that returns anti-k_t like jets, although two proposals have recently emerged [57, 58].

At the moment, quark and gluon jets can only be discriminated in a probabilistic fashion, i.e. constructing event samples that contain mostly quark or gluon jets [59]. The main strategy is to consider observables whose distributions are different for quark and gluon jets, and use this information in a similar way as for boosted object searches. In the simplest approach, one finds one or two variables, and imposes suitable cuts that will enrich event samples with quark or gluon jets. This requires having some theoretical understanding on the distributions in such variables. Alternatively, one can feed a larger number of variables to a neural network, which will then take care of finding the combination with the best discriminating power. One can have an even blinder approach, by just giving as input the momenta of all particles (alternatively their deposits in the detectors), and train a neural network using parton-shower event generators to perform a binary classification between quarks and gluon jets. Whatever approach one takes, the final aim is to construct a ROC curve displaying for instance the quark acceptance ε_{q} versus the gluon acceptance ε_{g}. If one is interested in tagging quark jets, the aim is to find a discriminant for which ε_{q} is large and ε_{g} is small. It is thus natural to exploit fact that gluons radiate roughly twice as much as quarks. Therefore, any observable that is sensitive to QCD radiation can be used for this purpose. The final-state variables that was first proposed and shown to have a good discriminating power was the girth of a jet J [60], defined as

$$g = \sum_{i \in \text{jet}} \frac{p_{ti}}{p_{tJ}} \Delta R_{iJ}, \tag{4.25}$$

where ΔR_{iJ} is the distance between the direction of particle p_i and the jet axis in the rapidity–azimuth plane. Note that this observable is closely related to the generalised 1-jettiness variable $\tau_1^{(1)}$, defined in equation (4.14). Before reviewing the performance of different observables, we discuss the basics of quark–gluon jet discrimination based on IRC safe jet observables. The observables that we consider here are zero for a single massless particle, and acquire a non-zero value dynamically through QCD radiation. As for the jet-mass distribution in figure 4.9, the girth distribution has a peak at a non-zero value of g. The position of the peak is different for quark and gluon jets, and at higher values of g for gluon jets than for quark jets. Quark/gluon jet discrimination can be then achieved imposing a cut the girth distribution at a value g_{cut} so as to include the peak of the girth distribution for quark jets, and leave out the peak of the same distribution for gluon jets. At leading logarithmic accuracy, the fraction of quark and gluon jets surviving the cut is just given by the two cumulative distributions

$$\Sigma_q(g_{\text{cut}}) \simeq \mathrm{e}^{-C_F \frac{\alpha_s}{\pi} \ln^2 g_{\text{cut}}}, \quad \Sigma_g(g_{\text{cut}}) \simeq \mathrm{e}^{-C_A \frac{\alpha_s}{\pi} \ln^2 g_{\text{cut}}}. \tag{4.26}$$

We first observe that, if $\Sigma_q(g_{\text{cut}}) = x$, i.e. one keeps a fraction x of quark jets, one automatically keeps a fraction x^{C_A/C_F} of gluon jets. This scaling holds at LL accuracy not only for the girth, but for any IRC safe final-state observable whose double logarithms exponentiate. This is due to the fact that, at LL level, both Σ_q and Σ_g are given by Sudakov exponents, proportional to the colour charge of the parton that has initiated the jet. This means that the different discriminating power of event shapes is due to subleading effects, which are generally not very large, and moreover are not treated in the same way by parton-shower event generators, which are the tools used in both theoretical and experimental studies of quark–gluon jet discrimination. In particular, beyond LL approximation, the discriminating power of an event shape depends on the relative importance of soft and collinear radiation. ECFs and their ratios have a built-in handle for this, their parameter β. For instance, one can consider [28] the ratio $C_1^{(\beta)}$ defined in equation (4.16). Note that, for $\beta = 1$, this variable has the same properties as the girth. Analytical studies of $\Sigma_{q/g}(C_1^{(\beta)})$ for different values of β suggest that smaller values of this parameter (e.g. $\beta = 0.2$) give a better quark–gluon jet discrimination, although the improvement with respect to the girth is not dramatic, being driven by subleading effects. Nevertheless, IRC safe jet observables have the advantage that the ROC output can be computed to a given perturbative accuracy. This is precisely what was done in the study of [28], where ROC curves were computed for the ratio of moments of ECFs $C_1^{(\beta)}$ (see equation (4.16)) corresponding to different values of β. The result is displayed in the left-hand panel of figure 4.22. The same behaviour is seen with Monte Carlo event generators, although numbers differ slightly due to differences in the treatment of subleading effects. At LL accuracy there is no dependence on β, because the ratio between the two efficiencies is determined by colour factors only. Differences start to appear at

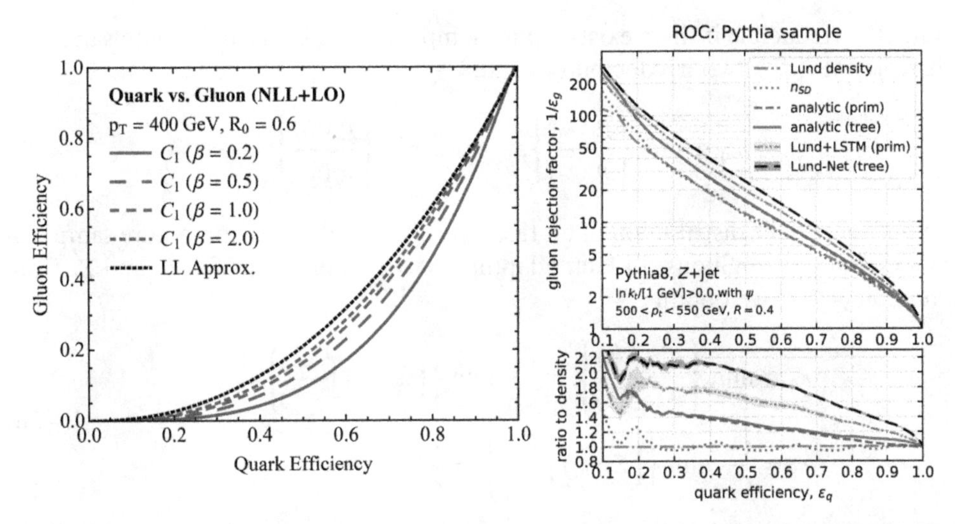

Figure 4.22. Left: gluon-jet tagging efficiency versus quark-jet tagging efficiency for the ratio of moments of ECFs $C_1^{(\beta)}$. Reproduced from [28] (2013). With permission of Springer. Right: ROC curve corresponding to different quark/gluon discrimination procedures utilising Lund-plane variables. © [61] (2022). With permission of Springer.

NLL, where one sees that increasing the importance of collinear versus soft contributions (i.e. decreasing β) increases the discriminatory power of an observable. Besides IRC safe observables, one can consider also unsafe ones, like charged hadron multiplicities, or jet electric charges [62]. In particular, the jet multiplicity seems to be the one with the best performance. To overcome the drawback of a lack of a perturbative description for this observables, alternatives have been devised with similar properties. One of those is the soft-drop multiplicity, which counts effectively the number of primary emissions in the Lund jet plane up to a given threshold [63]. This observable has been included in a larger set of observables constructed from the Lund jet plane [61]. All these variables are sensitive in many ways to the different QCD radiation patterns of quark and gluon jets. However, many of those have various degrees of correlation, so every observable-based discrimination procedure is nowadays fed into a neural network. An example of the ROC curve one can obtain from such methods can be found in the right-hand panel of figure 4.22, taken from [61]. There one can see that the neural network LundNet [64], using Lund-plane emission variables as inputs, outperforms all other discriminants, including the soft-drop multiplicity $n_{\rm SD}$ and the primary Lund jet plane density.

The main question in the discussion we have presented so far is what ultimately drives the performance of an observable with respect to the other. A concept that helps shed light on this issue is that of mutual irreducibility, introduced for quark and gluon jets in [65], but that can be applied to other contexts. Suppose we have a set of observables $\vec{x}$ (e.g. the value of the histograms of the invariant mass distribution of a jet), and we denote with $p_q(\vec{x})$ and $p_g(\vec{x})$ the probability that a jet is a quark or gluon jet respectively. The two distributions $p_q(\vec{x})$ and $p_g(\vec{x})$ are mutually irreducible if there exists a pure sample of quark jets and a pure sample of gluon jets, i.e. the two irreducibility factors

$$\kappa_{\rm q} \equiv \min_i \left\{ \frac{p_g(x_i)}{p_q(x_i)} \right\}, \qquad \kappa_{\rm g} \equiv \min_i \left\{ \frac{p_q(x_i)}{p_g(x_i)} \right\} \tag{4.27}$$

are both zero. The distribution of an IRC safe observable does give a pure sample of quarks, but not of gluons. In fact, adapting the LL expressions for Σ_q and Σ_g from equation (4.26), we obtain

$$\begin{aligned} \kappa_{\rm q} &= \min_i \left\{ \frac{d\Sigma_g/dx_i}{d\Sigma_q/dx_i} \right\} = \frac{C_{\rm A}}{C_{\rm F}} \min_i \left\{ \left[\Sigma_q(x_i)\right]^{\frac{C_{\rm A}}{C_{\rm F}}-1} \right\} = 0, \\ \kappa_{\rm g} &= \frac{C_{\rm F}}{C_{\rm A}} \min_i \left\{ \left[\Sigma_q(x_i)\right]^{1-\frac{C_{\rm A}}{C_{\rm F}}} \right\} = \frac{C_{\rm F}}{C_{\rm A}}. \end{aligned} \tag{4.28}$$

The physical interpretation is that, due to the Sudakov suppression and the fact that $C_{\rm A} - C_{\rm F} > 0$, for quarks one can always find a pure sample for $x \to 0$, but not for gluons. While we do not have access to $p_q(\vec{x})$ and $p_g(\vec{x})$, we can perform two different measurements $p_1(\vec{x})$ and $p_2(\vec{x})$, that depend on the set of observables $\vec{x}$, for instance the invariant mass of a jet in Z+jet or dijet events, and extract their irreducibility factors from data. In the assumption that quark and gluon jets in each sample are determined by the *same* probabilities $p_q(\vec{x})$ and $p_g(\vec{x})$, we have

$$p_i(\vec{x}) = f_i\, p_q(\vec{x}) + (1 - f_i)\, p_g(\vec{x}), \tag{4.29}$$

with f_i the fraction of events containing quark jets contributing to $p_i(\vec{x})$. The first observation is that, the likelihood ratio $p_1(\vec{x})/p_2(\vec{x})$ is a monotonic function of $p_q(\vec{x})/p_g(\vec{x})$. Therefore, optimising $p_1(\vec{x})/p_2(\vec{x})$ results in a powerful discriminant between quark and gluon jets, although it does not give yet access to $p_q(\vec{x})$ and $p_g(\vec{x})$. This is the basis of the neural-network discriminator CWoLa [66]. Second, if $\kappa_{\rm q} = \kappa_{\rm g} = 0$, $p_q(\vec{x})$ and $p_g(\vec{x})$ can be indeed extracted from $p_1(\vec{x})$ and $p_2(\vec{x})$ as follows:

$$p_q(\vec{x}) = \frac{p_1(\vec{x}) - \kappa_2\, p_2(\vec{x})}{1 - \kappa_2}, \qquad p_g(\vec{x}) = \frac{p_2(\vec{x}) - \kappa_1 p_1(\vec{x})}{1 - \kappa_1}, \tag{4.30}$$

where κ_1 and κ_2 are the irreducibility factors for $p_1(\vec{x})$ and $p_2(\vec{x})$, defined as in equation (4.27). Given $p_1(\vec{x})$ and $p_2(\vec{x})$, it is possible to find two mutually irreducible distributions $p_{T_1}(\vec{x})$ and $p_{T_2}(\vec{x})$, corresponding to two 'topics' T_1 and T_2, in the jargon used in language processing. In general, these distributions are related to $p_q(\vec{x})$ and $p_g(\vec{x})$ by

$$p_{T_1} = \frac{p_q(\vec{x}) - \kappa_{\rm g}\, p_g(\vec{x})}{1 - \kappa_{\rm g}}, \qquad p_{T_2} = \frac{p_g(\vec{x}) - \kappa_{\rm q}\, p_q(\vec{x})}{1 - \kappa_{\rm q}}. \tag{4.31}$$

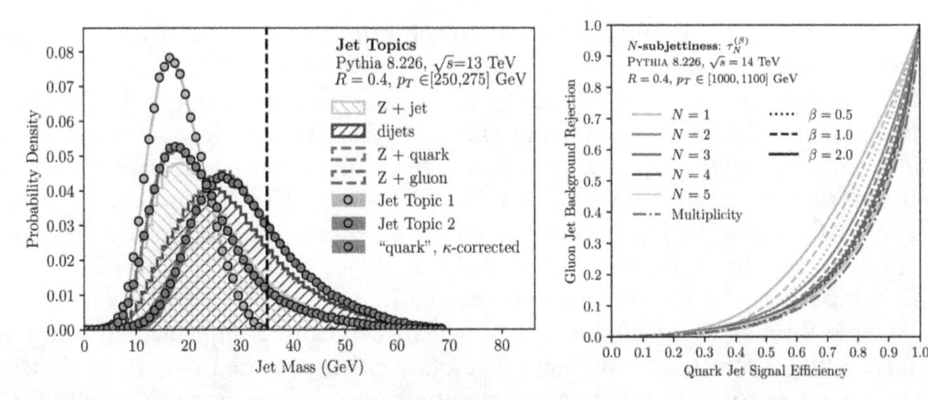

Figure 4.23. Left: the distribution in the invariant mass of a jet in Z+jet events, and its flavour and topic decompositions. The details of the simulation can be found in the figure [65]. Reprinted (figure 4) with permission from [65], Copyright (2018) by the American Physical Society. Right: The ROC curve for N-subjettiness and constituent multiplicity for simulated jet events. © [67] (2019). With permission of Springer.

Therefore, if $\kappa_{\mathrm{q}} = \kappa_{\mathrm{g}} = 0$, which is obtained by using the particle multiplicity in a jet as an observable, we obtain $p_{T_1}(\vec{x}) = p_q(\vec{x})$ and $p_{T_2}(\vec{x}) = p_g(\vec{x})$. This is the main finding of [65]. It is interesting to observe what happens using the mass distribution, where $\kappa_{\mathrm{q}} \simeq 0$ and $\kappa_{\mathrm{g}} \neq 0$. There we expect that the distribution in T_2 will yield the gluon jet probability, whereas the one in T_1 will be a superposition of $p_q(\vec{x})$ and $p_g(\vec{x})$. This is what is seen in the left-hand panel of figure 4.23. There, the distribution in the invariant mass of a jet in simulated Z+jet events is decomposed in two topics. Once one corrects the distribution of the first topic with the gluon-jet irreducibility factor obtained from the parton-shower event generator, one correctly reproduces the mass distribution for Z plus quark jets.

A last important finding we wish to present is the relation between IRC safe observables and the multiplicity [67]. Reference [67] considers the quark/gluon discriminating power of N-subjettiness, for increasing values of N. One interesting result is that, while in all cases $\kappa_{\mathrm{q}} = 0$, κ_{g} is equal to $(C_{\mathrm{F}}/C_{\mathrm{A}})^N$. This implies that, for large N, N-subjettiness is expected to perform in a similar way to multiplicity. Physically, this means that, in order to be able to select a pure gluon sample, one needs to have information on all emissions. This finding is confirmed in the left-hand panel of figure 4.23, taken from [67], which shows the ROC curve for N-subjettiness approaching that of the multiplicity for increasing N.

4.3.2 Quark- and gluon-jet discrimination at the LHC

Systematic studies of the properties of quark and gluon jets have been performed at LEP (see e.g. [68, 69]). These analyses take advantage of three-jet events in which two b-tagged jets fall in the same hemisphere, thus leaving an energetic gluon jet in the other hemisphere to obtain pure samples of gluon jets. This information is used to measure the particle multiplicities inside quark and gluon jets, and extract from that the ratio $C_{\mathrm{A}}/C_{\mathrm{F}}$. In fact, the ratio of hadron multiplicity in quark and gluon jets tends, for asymptotically high energies, to the ratio of the corresponding colour factors. A measurement of C_A/C_F has also been performed at the Tevatron [70].

LHC experiments have exploited the ideas illustrated in section 4.3.1 to build statistical discriminants between quark and gluon jets. An overview of the available results can be found in [45]. Instead of listing the results presented there, we wish to focus one two recent analyses, one by CMS and one by ATLAS at $\sqrt{s} = 13$ TeV.

The CMS analysis of [71] presents measurements of angularities, defined in equation (4.19), in quark-enriched sample in Z+jet events. The enrichment is achieved using parton-shower event generators. In particular, the analysis identifies the sample through event selection cuts, which then can be implemented into the corresponding theoretical predictions. These are either provided by parton-shower event generators, or even analytic resummations at NLL accuracy (labelled NLO +NLL'+NP in the figure). In particular, figure 4.24 shows the IRC safe ($\kappa = 1$) angularity distributions corresponding to the angular exponent $\beta = 1/2$ and (see equation (4.19)), which enhances the importance of the collinear region with respect to the soft one. In particular, the left-hand panel corresponds to ungroomed jets, whereas the right-hand to groomed jets with the soft-drop procedure, with parameters $z_{\text{cut}} = 0.1$ and $\beta = 0$. There one sees quite a good agreement with parton-shower event generators, as well as fixed-order predictions, which are appropriate in the tail of distributions. The agreement is poorer for resummation close to the peaks of distribution. The experimental accuracy is remarkable, which calls already for better logarithmic accuracy in the resummation.

Another study performed by ATLAS [72] investigates the possibility of using the image of a jet in the rapidity–azimuth plane as an input to a convolutional neural

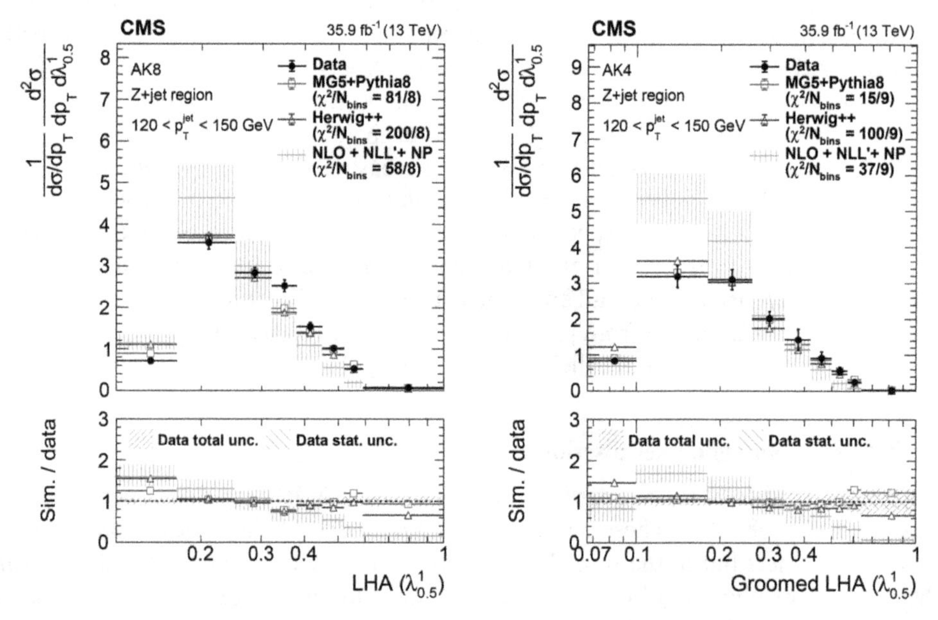

Figure 4.24. Experimental measurements of an angularity [71] for ungroomed (left) and groomed (right) jets, compared to various theoretical predictions. © [71] (2022). With permission of Springer.

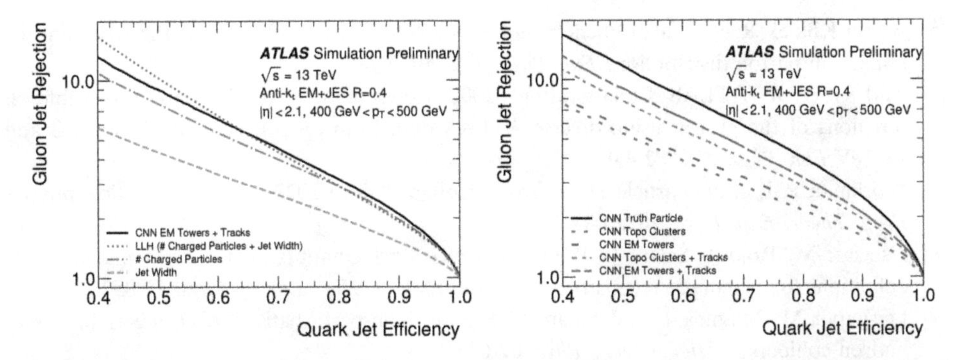

Figure 4.25. Left: ROC curves corresponding to different procedures to discriminate between quark and gluon jets. Right: ROC curves corresponding to a CNN based on jet images with different inputs. Both plots are taken from reference [72], ATLAS Experiment, Copyright 2017 CERN.

network (CNN) as a means to distinguish between quark and gluon jets via image recognition. The presented results correspond to simulations performed with parton-shower event generators. The left-hand panel of figure 4.25 shows the ROC curve for the inverse of gluon jet efficiency (labelled gluon jet rejection) versus quark-jet efficiency, corresponding to different procedures to discriminate quark and gluon jets. We notice that the CNN outperforms discrimination based solely on IRC safe observables (the jet width, essentially the girth in equation (4.25)), as well as solely on the tracked particle multiplicity. However, when the two are combined using a two-dimensional likelihood, one obtains comparable results with respect to the CNN. The other important aspect to consider is the dependence of the performance of the CNN with respect to its input. This is shown in the right-hand panel of figure 4.25, where we see important differences on whether one uses tracks only, adds calorimeter information, or uses full particle information as available from the parton-shower event generators. This points to a very important question that every tagger has to address, which is that of its robustness with respect to its input. Another important aspect is the resilience of the tagger against less controlled effects such as hadronisation and underlying events. These are two relevant issues that all jet-substructure studies will have to confront with in the future.

References

[1] Aaboud M *et al* (ATLAS Collaboration) 2018 Measurement of inclusive jet and dijet cross-sections in proton–proton collisions at $\sqrt{s} = 13$ TeV with the ATLAS detector *J. High Energy Phys.* **05** 195

[2] Nagy Z 2002 Three jet cross-sections in hadron hadron collisions at next-to-leading order *Phys. Rev. Lett.* **88** 122003

[3] Khachatryan V *et al* (CMS Collaboration) 2017 Measurement and QCD analysis of double-differential inclusive jet cross sections in pp collisions at $\sqrt{s} = 8$ TeV and cross section ratios to 2.76 and 7 TeV *J. High Energy Phys.* **03** 156

[4] Currie J, Glover E W N and Pires J 2017 Next-to-next-to leading order QCD predictions for single jet inclusive production at the LHC *Phys. Rev. Lett.* **118** 072002

[5] Abdul Khalek R *et al* 2020 Phenomenology of NNLO jet production at the LHC and its impact on parton distributions *Eur. Phys. J.* C **80** 797
[6] Aad G *et al* (ATLAS Collaboration) 2002 Determination of the parton distribution functions of the proton using diverse ATLAS data from *pp* collisions at $\sqrt{s} = 7$, 8 and 13 TeV *Eur. Phys. J.* C **82** 438
[7] Workman R L *et al* (Particle Data Group Collaboration) 2022 Review of particle physics *Prog. Theor. Exp. Phys.* **2022** 083C01
[8] Cacciari M, Rojo J, Salam G P and Soyez G 2008 Quantifying the performance of jet definitions for kinematic reconstruction at the LHC *J. High Energy Phys.* **12** 032
[9] Dasgupta M, Magnea L and Salam G P 2008 Non-perturbative QCD effects in jets at hadron colliders *J. High Energy Phys.* **02** 055
[10] Dasgupta M, Dreyer F, Salam G P and Soyez G 2015 Small-radius jets to all orders in QCD *J. High Energy Phys.* **04** 039
[11] Ellis S D, Huston J, Hatakeyama K, Loch P and Tonnesmann M 2008 Jets in hadron-hadron collisions *Prog. Part. Nucl. Phys.* **60** 484–551
[12] Seymour M H 1994 Searches for new particles using cone and cluster jet algorithms: a comparative study *Z. Phys.* C **62** 127–38
[13] Larkoski A J, Moult I and Nachman B 2020 Jet substructure at the large hadron collider: a review of recent advances in theory and machine learning *Phys. Rep.* **841** 1–63
[14] Kaplan D E, Rehermann K, Schwartz M D and Tweedie B 2008 Top tagging: a method for identifying boosted hadronically decaying top quarks *Phys. Rev. Lett.* **101** 142001
[15] Plehn T, Salam G P and Spannowsky M 2010 Fat jets for a light Higgs *Phys. Rev. Lett.* **104** 111801
[16] Plehn T, Spannowsky M, Takeuchi M and Zerwas D 2010 Stop reconstruction with tagged tops *J. High Energy Phys.* **10** 078
[17] Schaetzel S and Spannowsky M 2014 Tagging highly boosted top quarks *Phys. Rev.* D **89** 014007
[18] Krohn D, Thaler J and Wang L-T 2010 Jet trimming *J. High Energy Phys.* **02** 084
[19] Ellis S D, Vermilion C K and Walsh J R 2010 Recombination algorithms and jet substructure: pruning as a tool for heavy particle searches *Phys. Rev.* D **81** 094023
[20] Butterworth J M, Davison A R, Rubin M and Salam G P 2008 Jet substructure as a new Higgs search channel at the LHC *Phys. Rev. Lett.* **100** 242001
[21] Larkoski A J, Marzani S, Soyez G and Thaler J 2014 Soft drop *J. High Energy Phys.* **05** 146
[22] Dasgupta M, Fregoso A, Marzani S and Salam G P 2013 Towards an understanding of jet substructure *J. High Energy Phys.* **09** 029
[23] Rubin M 2010 Non-global logarithms in filtered jet algorithms *J. High Energy Phys.* **05** 005
[24] Butterworth J M, Cox B E and Forshaw J R 2002 WW scattering at the CERN LHC *Phys. Rev.* D **65** 096014
[25] Thaler J and Van Tilburg K 2011 Identifying boosted objects with N-subjettiness *J. High Energy Phys.* **03** 015
[26] Thaler J and Van Tilburg K 2012 Maximizing boosted top identification by minimizing N-subjettiness *J. High Energy Phys.* **02** 093
[27] Jouttenus T T, Stewart I W, Tackmann F J and Waalewijn W J 2011 The soft function for exclusive N-jet production at hadron colliders *Phys. Rev.* D **83** 114030
[28] Larkoski A J, Salam G P and Thaler J 2013 Energy correlation functions for jet substructure *J. High Energy Phys.* **06** 108

[29] Larkoski A J, Moult I and Neill D 2016 Analytic boosted boson discrimination *J. High Energy Phys.* **05** 117
[30] Komiske P T, Metodiev E M and Thaler J 2018 Energy flow polynomials: a complete linear basis for jet substructure *J. High Energy Phys.* **04** 013
[31] Larkoski A J, Thaler J and Waalewijn W J 2014 Gaining (mutual) information about quark/gluon discrimination *J. High Energy Phys.* **11** 129
[32] Gallicchio J, Huth J, Kagan M, Schwartz M D, Black K and Tweedie B 2011 Multivariate discrimination and the Higgs + W/Z search *J. High Energy Phys.* **04** 069
[33] Qu H and Gouskos L 2020 ParticleNet: jet tagging via particle clouds *Phys. Rev.* D **101** 056019
[34] Frye C, Larkoski A J, Schwartz M D and Yan K 2016 Precision physics with pile-up insensitive observables 1603.06375 [hep-ph]
[35] Marzani S, Schunk L and Soyez G 2018 The jet mass distribution after soft drop *Eur. Phys. J.* C **78** 96
[36] Kang Z-B, Lee K, Liu X and Ringer F 2018 The groomed and ungroomed jet mass distribution for inclusive jet production at the LHC *J. High Energy Phys.* **10** 137
[37] Gallicchio J and Schwartz M D 2010 Seeing in color: jet superstructure *Phys. Rev. Lett.* **105** 022001
[38] Larkoski A, Marzani S and Wu C 2020 Safe use of jet pull *J. High Energy Phys.* **01** 104
[39] Cogan J, Kagan M, Strauss E and Schwarztman A 2015 Jet-images: computer vision inspired techniques for jet tagging *J. High Energy Phys.* **02** 118
[40] de Oliveira L, Kagan M, Mackey L, Nachman B and Schwartzman A 2016 Jet-images–deep learning edition *J. High Energy Phys.* **07** 069
[41] Andersson B, Gustafson G, Lonnblad L and Pettersson U 1989 Coherence effects in deep inelastic scattering *Z. Phys.* C **43** 625
[42] Dreyer F A, Salam G P and Soyez G 2018 The Lund jet plane *J. High Energy Phys.* **12** 064
[43] Ellis S D, Hornig A, Roy T S, Krohn D and Schwartz M D 2012 Qjets: a non-deterministic approach to tree-based jet substructure *Phys. Rev. Lett.* **108** 182003
[44] Kahawala D, Krohn D and Schwartz M D 2013 Jet sampling: improving event reconstruction through multiple interpretations *J. High Energy Phys.* **06** 006
[45] Kogler R *et al* 2019 Jet substructure at the large hadron collider: experimental review *Rev. Mod. Phys.* **91** 045003
[46] Aaboud M *et al* (ATLAS Collaboration) 2019 Performance of top-quark and W-boson tagging with ATLAS in Run 2 of the LHC *Eur. Phys. J.* C **79** 375
[47] Aaboud M *et al* (ATLAS Collaboration) 2018 Measurement of colour flow using jet-pull observables in $t\bar{t}$ events with the ATLAS experiment at $\sqrt{s} = 13$ TeV *Eur. Phys. J.* C **78** 847
[48] Aad G *et al* (ATLAS Collaboration) 2020 Measurement of soft-drop jet observables in pp collisions with the ATLAS detector at $\sqrt{s} = 13$ TeV *Phys. Rev.* D **101** 052007
[49] Kang Z-B, Lee K, Liu X, Neill D and Ringer F 2020 The soft drop groomed jet radius at NLL *J. High Energy Phys.* **02** 054
[50] Sirunyan A M *et al* (CMS Collaboration) 2018 *J. High Energy Phys.* **11** 113
[51] Aad G *et al* (ATLAS Collaboration) 2020 *Phys. Rev. Lett.* **124** 222002
[52] Tumasyan A *et al* (CMS Collaboration) 2022 Search for resonances decaying to three W bosons in proton-proton collisions at $\sqrt{s} = 13$ TeV *Phys. Rev. Lett.* **129** 021802

[53] Sirunyan A M *et al* (CMS Collaboration) 2020 Identification of heavy, energetic, hadronically decaying particles using machine-learning techniques *J. Instrum.* **15** P06005

[54] Adam W and Vivarelli I 2022 Status of searches for electroweak-scale supersymmetry after LHC Run 2 *Int. J. Mod. Phys.* A **37** 2130022

[55] Banfi A, Salam G P and Zanderighi G 2006 Infrared safe definition of jet flavor *Eur. Phys. J.* C **47** 113–24

[56] Banfi A, Salam G P and Zanderighi G 2007 Accurate QCD predictions for heavy-quark jets at the Tevatron and LHC *J. High Energy Phys.* **07** 026

[57] Czakon M, Mitov A and Poncelet R 2022 Infrared-safe flavoured anti-k_T jets Arxiv:2205.11879 [hep-ph]

[58] Gauld R, Huss A and Stagnitto G 2022 A dress of flavour to suit any jet Arxiv:2208.11138 [hep-ph]

[59] Gras P, Höche S, Kar D, Larkoski A, Lönnblad L, Plätzer S, Siódmok A, Skands P, Soyez G and Thaler J 2017 Systematics of quark/gluon tagging *J. High Energy Phys.* **07** 091

[60] Gallicchio J and Schwartz M D 2011 Quark and gluon tagging at the LHC *Phys. Rev. Lett.* **107** 172001

[61] Dreyer F A, Soyez G and Takacs A 2022 Quarks and gluons in the Lund plane *J. High Energy Phys.* **08** 177

[62] Krohn D, Schwartz M D, Lin T and Waalewijn W J 2013 Jet charge at the LHC *Phys. Rev. Lett.* **110** 212001

[63] Frye C, Larkoski A J, Thaler J and Zhou K 2017 Casimir meets Poisson: improved quark/gluon discrimination with counting observables *J. High Energy Phys.* **09** 083

[64] Dreyer F A and Qu H 2021 Jet tagging in the Lund plane with graph networks *J. High Energy Phys.* **03** 052

[65] Metodiev E M and Thaler J 2018 Jet topics: disentangling quarks and gluons at colliders *Phys. Rev. Lett.* **120** 241602

[66] Metodiev E M, Nachman B and Thaler J 2017 Classification without labels: learning from mixed samples in high energy physics *J. High Energy Phys.* **10** 174

[67] Larkoski A J and Metodiev E M 2019 A theory of quark versus gluon discrimination *J. High Energy Phys.* **10** 014

[68] Abbiendi G *et al* (OPAL Collaboration) 1999 Experimental properties of gluon and quark jets from a point source *Eur. Phys. J.* C **11** 217–38

[69] Abreu P *et al* (DELPHI Collaboration) 1999 The scale dependence of the hadron multiplicity in quark and gluon jets and a precise determination of C_A/C_F *Phys. Lett.* B **449** 383–400

[70] Pronko A (CDF Collaboration) 2005 Fragmentation differences of quark and gluon jets at Tevatron *Int. J. Mod. Phys.* A **20** 3723–5

[71] Tumasyan A *et al* (CMS Collaboration) 2022 Study of quark and gluon jet substructure in Z+jet and dijet events from pp collisions *J. High Energy Phys.* **01** 188

[72] ATLAS Collaboration, Quark versus gluon jet tagging using jet images with the ATLAS detector http://cds.cern.ch/record/2275641

CPSIA information can be obtained
at www.ICGtesting.com
Printed in the USA
BVHW091402230123
656820BV00004B/83